# ELEMENTARY
# CLASSICAL
# MECHANICS
## Problems and Solutions

# ELEMENTARY CLASSICAL MECHANICS
## Problems and Solutions

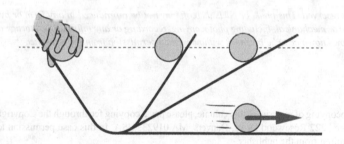

## Stephen Wiggins
### University of Bristol, UK

**W World Scientific**

NEW JERSEY · LONDON · SINGAPORE · BEIJING · SHANGHAI · HONG KONG · TAIPEI · CHENNAI · TOKYO

*Published by*

World Scientific Publishing Co. Pte. Ltd.

5 Toh Tuck Link, Singapore 596224

*USA office:* 27 Warren Street, Suite 401-402, Hackensack, NJ 07601

*UK office:* 57 Shelton Street, Covent Garden, London WC2H 9HE

Library of Congress Control Number: 2023026153

**British Library Cataloguing-in-Publication Data**
A catalogue record for this book is available from the British Library.

**ELEMENTARY CLASSICAL MECHANICS**
**Problems and Solutions**

ISBN 978-981-127-748-1 (paperback)
ISBN 978-981-127-749-8 (ebook for institutions)
ISBN 978-981-127-750-4 (ebook for individuals)

For any available supplementary material, please visit
https://www.worldscientific.com/worldscibooks/10.1142/13444#t=suppl

Typeset by Stallion Press
Email: enquiries@stallionpress.com

Printed in Singapore

# Preface

This book contains the solutions to all of the problems in my book *Elementary Classical Mechanics*. While both books originated during my time in the city of Bristol, United Kingdom, I have completed it in the past few months during my stay at the United States Naval Academy in Annapolis, Maryland. I am grateful for the support of the William R. Davis '08 Chair in the Department of Mathematics at the United States Naval Academy.

# Contents

# List of Figures

# Chapter 1

# Solutions for Problem Set 1

**Problem 1.** For the vectors shown in the figure below construct:

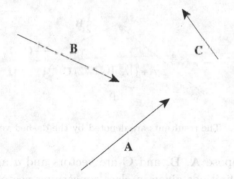

(a) $-\mathbf{A} + \mathbf{B} + 2\mathbf{C}$,

(b) $\mathbf{A} - \mathbf{B} - 2\mathbf{C}$,

(c) $2\mathbf{A} + \mathbf{B} - \mathbf{C}$,

(d) $\mathbf{A} + \frac{1}{2}\mathbf{B} - \frac{1}{2}\mathbf{C}$.

What is the relation between the vectors you constructed in (a) and (b)?
What is the relation between the vectors you constructed in (c) and (d)?

**Solution.** The graphical constructions are given in Fig. 1.1. The resultants in (a) and (b) have equal length and opposite direction. The resultants in (c) and (d) have the same direction, but the resultant in (c) has twice the length of the resultant in (d).

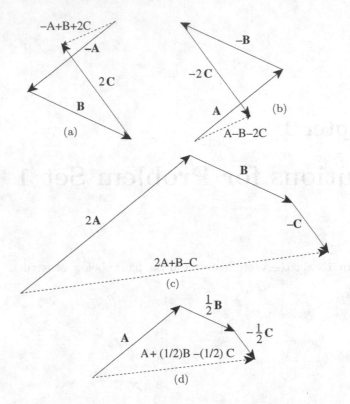

Fig. 1.1. The resultants are denoted by the dashed vectors.

**Problem 2.** Suppose **A**, **B**, and **C** are vectors and $a$ and $b$ are scalars. Using only the definitions given in class concerning vectors, and the rules you already know for manipulating scalars, graphically demonstrate the following laws of vector algebra.

(a) $\mathbf{A} + \mathbf{B} = \mathbf{B} + \mathbf{A}$,    Commutative Law for Vector Addition,
(b) $\mathbf{A} + (\mathbf{B} + \mathbf{C}) = (\mathbf{A} + \mathbf{B}) + \mathbf{C}$,    Associative Law for Vector Addition,
(c) $a(b\mathbf{A}) = (ab)\mathbf{A} = b(a\mathbf{A})$,    Associative Law for Scalar Multiplication,
(d) $(a + b)\mathbf{A} = a\mathbf{A} + b\mathbf{A}$,    Distributive Law,
(e) $a(\mathbf{A} + \mathbf{B}) = a\mathbf{A} + a\mathbf{B}$,    Distributive Law.

**Solution.** The same vectors from Fig. 1.1 of the problems are used here.

(a) This is shown in Fig. 1.2.
(b) Referring to Fig. 1.3, in (i) $\mathbf{A} + \mathbf{B}$ is constructed, and in (ii) $\mathbf{C}$ is added to $\mathbf{A} + \mathbf{B}$. In (iii) $\mathbf{B} + \mathbf{C}$ is constructed and in (iv) $\mathbf{B} + \mathbf{C}$ is added to **A**. It follows that (ii) = (iv).

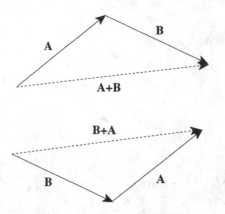

Fig. 1.2. **A** + **B** = **B** + **A**.

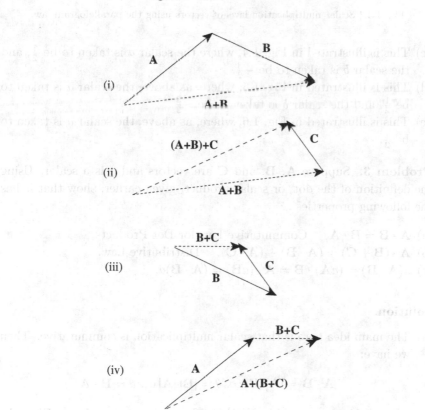

Fig. 1.3. Addition and subtraction laws of vectors using the parallelogram law.

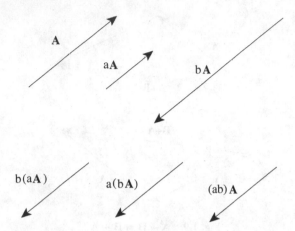

Fig. 1.4. Scalar multiplication laws of vectors using the parallelogram law.

(c) This is illustrated in Fig. 1.4, where the scalar $a$ is taken to be $\frac{1}{2}$, and the scalar $b$ is taken to be $-\frac{3}{2}$.

(d) This is illustrated in Fig. 1.5, where, as above, the scalar $a$ is taken to be $\frac{1}{2}$, and the scalar $b$ is taken to be $-\frac{3}{2}$.

(e) This is illustrated in Fig. 1.6, where, as above, the scalar $a$ is taken to be $\frac{1}{2}$.

**Problem 3.** Suppose $\mathbf{A}$, $\mathbf{B}$, and $\mathbf{C}$ are vectors and $a$ is a scalar. Using the definition of the dot, or scalar product, given earlier, show that it has the following properties.

(a) $\mathbf{A} \cdot \mathbf{B} = \mathbf{B} \cdot \mathbf{A}$,    Commutative Law for Dot Products,

(b) $\mathbf{A} \cdot (\mathbf{B} + \mathbf{C}) = (\mathbf{A} \cdot \mathbf{B}) + (\mathbf{A} \cdot \mathbf{C})$,    Distributive Law,

(c) $a(\mathbf{A} \cdot \mathbf{B}) = (a\mathbf{A}) \cdot \mathbf{B} = \mathbf{A} \cdot (a\mathbf{B}) = (\mathbf{A} \cdot \mathbf{B})a$.

**Solution.**

(a) The main idea here is that scalar multiplication is commutative. Then we have:

$$\mathbf{A} \cdot \mathbf{B} = |\mathbf{A}|\,|\mathbf{B}| \cos\theta = |\mathbf{B}|\,|\mathbf{A}| \cos\theta = \mathbf{B} \cdot \mathbf{A},$$

where it is assumed evidently that "the angle between $\mathbf{A}$ and $\mathbf{B}$" is the same as the "the angle between $\mathbf{B}$ and $\mathbf{A}$".

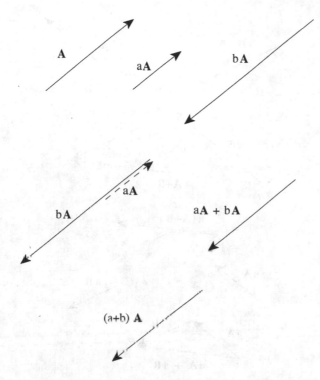

Fig. 1.5. Scalar multiplication laws of vectors using the parallelogram law.

(b) In order to prove this we use the idea of the projection of one vector on another. We will take it as evident that the projection of $\mathbf{B} + \mathbf{C}$ on $\mathbf{A}$ is the projection of $\mathbf{B}$ on $\mathbf{A}$ plus the projection of $\mathbf{C}$ on $\mathbf{A}$. Let $\mathbf{a} \equiv \frac{\mathbf{A}}{|\mathbf{A}|}$. Then the previous statement is mathematically written as:

$$(\mathbf{B} + \mathbf{C}) \cdot \mathbf{a} = \mathbf{B} \cdot \mathbf{a} + \mathbf{C} \cdot \mathbf{a}.$$

Multiplying this expression by $|\mathbf{A}|$ gives:

$$(\mathbf{B} + \mathbf{C}) \cdot |\mathbf{A}|\mathbf{a} = \mathbf{B} \cdot |\mathbf{A}|\mathbf{a} + \mathbf{C} \cdot |\mathbf{A}|\mathbf{a},$$

or

$$(\mathbf{B} + \mathbf{C}) \cdot \mathbf{A} = \mathbf{B} \cdot \mathbf{A} + \mathbf{C} \cdot \mathbf{A}.$$

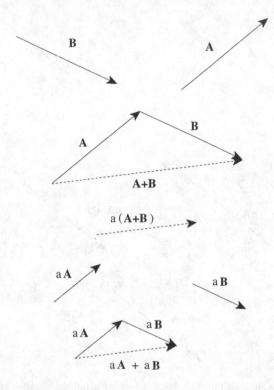

Fig. 1.6. Scalar multiplication and vector addition laws using the parallelogram law.

Using commutativity of the dot product gives:

$$\mathbf{A} \cdot (\mathbf{B} + \mathbf{C}) = \mathbf{B} \cdot \mathbf{A} + \mathbf{C} \cdot \mathbf{A}.$$

(c) The main idea here is that scalar multiplication is commutative. First, if $a \geq 0$ the result follows easily from applying the definition of the dot product. From the figure on the next page, it follows that:

$$-\mathbf{A} \cdot \mathbf{B} = -|\mathbf{A}| \, |\mathbf{B}| \cos \theta.$$

Using this result, the result obtained for $a > 0$ can be easily obtained for $a < 0$ following exactly the same reasoning.

**Problem 4.** Let $\mathbf{A}$ and $\mathbf{B}$ be nonzero vectors. Prove that $\mathbf{A} + \mathbf{B}$ and $\mathbf{A} - \mathbf{B}$ are perperpendicular if and only if $|\mathbf{A}| = |\mathbf{B}|$.

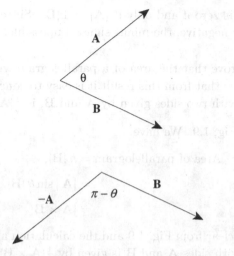

Fig. 1.7. Graphical illustration of the algebra of the dot product.

Draw a diagram to show that, geometrically, this means that if **A** and **B** are the sides of a parallelogram, then the parallelogram is, in fact, a rhombus.[1]

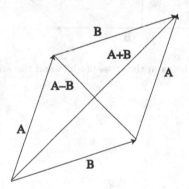

Fig. 1.8. Geometry of vector addition and subtraction and the relationship to a parallelogram.

**Solution.** We have
$$(\mathbf{A} + \mathbf{B}) \cdot (\mathbf{A} - \mathbf{B}) = \mathbf{A} \cdot \mathbf{A} - \mathbf{A} \cdot \mathbf{B} + \mathbf{B} \cdot \mathbf{A} - \mathbf{B} \cdot \mathbf{B} = |\mathbf{A}|^2 - |\mathbf{B}|^2.$$

---

[1]This is an important problem for understanding the geometric meaning of the dot product. Note that it does *not* require the use of any coordinate system.

This expression is zero if and only if $|\mathbf{A}| = \pm|\mathbf{B}|$. Since the magnitude of a vector is never negative, the minus sign is impossible.

**Problem 5.** Prove that the area of a parallelogram with sides $\mathbf{A}$ and $\mathbf{B}$ is $|\mathbf{A} \times \mathbf{B}|$. Argue that from this result it is easy to conclude that the area of the *triangle*, with two sides given by $\mathbf{A}$ and $\mathbf{B}$, is $\frac{1}{2}|\mathbf{A} \times \mathbf{B}|$.[2]

**Solution.** See Fig. 1.9. We have:

$$\text{Area of parallelogram} = h\,|\mathbf{B}|,$$
$$= |\mathbf{A}|\sin\theta\,|\mathbf{B}|,$$
$$= |\mathbf{A} \times \mathbf{B}|$$

It should be clear from Fig. 1.9 and the calculation above that the area of the triangle with sides $\mathbf{A}$ and $\mathbf{B}$ is given by $\frac{1}{2}|\mathbf{A} \times \mathbf{B}|$.

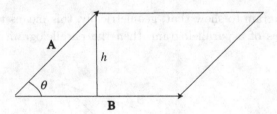

Fig. 1.9.   Relationship between the cross-product and the area of a parallelogram.

---

[2]Similar to the previous problem, this is an important problem for understanding the geometric meaning of the cross product. It also does *not* require the use of any coordinate system.

# Chapter 2

# Solutions for Problem Set 2

**Problem 1.** Suppose $\mathbf{A}$, $\mathbf{B}$, $\mathbf{C}$, and $\mathbf{D}$ are vectors and $a$ is a scalar. Using the definition of the cross, or vector product, given in class, *and expressing the vectors in rectangular coordinates*, show that the following properties hold.

(a) $\mathbf{A} \times \mathbf{B} = -\mathbf{B} \times \mathbf{A}$,   Commutative Law for Cross-Products Fails,
(b) $(\mathbf{A} + \mathbf{B}) \times (\mathbf{C} + \mathbf{D}) = (\mathbf{A} \times \mathbf{C}) + (\mathbf{B} \times \mathbf{C}) + (\mathbf{A} \times \mathbf{D}) + (\mathbf{B} \times \mathbf{D})$,
   Distributive Law,
(c) $a(\mathbf{A} \times \mathbf{B}) = (a\mathbf{A}) \times \mathbf{B} = \mathbf{A} \times (a\mathbf{B}) = (\mathbf{A} \times \mathbf{B})a$.

**Solution.** Let

$$\mathbf{A} = A_1\mathbf{i} + A_2\mathbf{j} + A_3\mathbf{k},$$

$$\mathbf{B} = B_1\mathbf{i} + B_2\mathbf{j} + B_3\mathbf{k},$$

$$\mathbf{C} = C_1\mathbf{i} + C_2\mathbf{j} + C_3\mathbf{k},$$

$$\mathbf{D} = D_1\mathbf{i} + D_2\mathbf{j} + D_3\mathbf{k}.$$

(a) We have

$$\mathbf{A} \times \mathbf{B} = (A_2B_3 - A_3B_2)\mathbf{i} + (A_3B_1 - A_1B_3)\mathbf{j} + (A_1B_2 - A_2B_1)\mathbf{k},$$

and

$$\mathbf{B} \times \mathbf{A} = (B_2A_3 - B_3A_2)\mathbf{i} + (B_3A_1 - B_1A_3)\mathbf{j} + (B_1A_2 - B_2A_1)\mathbf{k}.$$

By inspection, you see that $\mathbf{A} \times \mathbf{B} = -\mathbf{B} \times \mathbf{A}$.

9

(b) We have:

$$(\mathbf{A} + \mathbf{B}) \times (\mathbf{C} + \mathbf{D})$$
$$= ((A_2 + B_2)(C_3 + D_3) - (A_3 + B_3)(C_2 + D_2)) \, \mathbf{i}$$
$$+ ((A_3 + B_3)(C_1 + D_1) - (A_1 + B_1)(C_3 + D_3)) \, \mathbf{j}$$
$$+ ((A_1 + B_1)(C_2 + D_2) - (A_2 + B_2)(C_1 + D_1)) \, \mathbf{k}.$$

Also,

$$\mathbf{A} \times \mathbf{C} = (A_2 C_3 - A_3 C_2)\mathbf{i} + (A_3 C_1 - A_1 C_3)\mathbf{j} + (A_1 C_2 - A_2 C_1)\mathbf{k},$$

$$\mathbf{B} \times \mathbf{C} = (B_2 C_3 - B_3 C_2)\mathbf{i} + (B_3 C_1 - B_1 C_3)\mathbf{j} + (B_1 C_2 - B_2 C_1)\mathbf{k}.$$

$$\mathbf{A} \times \mathbf{D} = (A_2 D_3 - A_3 D_2)\mathbf{i} + (A_3 D_1 - A_1 D_3)\mathbf{j} + (A_1 D_2 - A_2 D_1)\mathbf{k},$$

$$\mathbf{B} \times \mathbf{D} = (B_2 D_3 - B_3 D_2)\mathbf{i} + (B_3 D_1 - B_1 D_3)\mathbf{j} + (B_1 D_2 - D_2 C_1)\mathbf{k}.$$

By inspection of these quantities it follows that the equality holds.

(c) Once $\mathbf{A} \times \mathbf{B}$ has been computed in components, the result becomes trivial as it just involves commutation of scalar multiplication.

**Problem 2.** Let $\mathbf{A} = 4\mathbf{i} + 3\mathbf{j} + 2\mathbf{k}$ and $\mathbf{B} = -\mathbf{i} + 7\mathbf{j} - 3\mathbf{k}$. Compute $\mathbf{A} \times \mathbf{B}$ and $\mathbf{A} \cdot \mathbf{B}$.

**Solution.** $\mathbf{A} \times \mathbf{B} = -23\mathbf{i} + 10\mathbf{j} + 31\mathbf{k}$, $\mathbf{A} \cdot \mathbf{B} = 11$.

**Problem 3.** Let $\mathbf{A} = 2\mathbf{i} + 4\mathbf{j} + 2\mathbf{k}$ and $\mathbf{B} = -4\mathbf{i} - 8\mathbf{j} - 4\mathbf{k}$. Determine $\mathbf{A} \times \mathbf{B}$ *without doing any computations*.

**Solution.** The two vectors are parallel (one is a scalar multiple of the other), so $\mathbf{A} \times \mathbf{B} = 0$.

**Problem 4.** Consider each of the expressions below. Which expressions "make sense" (i.e. have mathematical validity in terms of the properties of vectors). For those that do, use $\mathbf{a} = \mathbf{i} - \mathbf{j}$, $\mathbf{b} = \mathbf{i}$, $\mathbf{c} = \mathbf{j}$, and $\mathbf{d} = \mathbf{i} + \mathbf{j}$ to evaluate the expression.[1]

(a) $\frac{\mathbf{a} \cdot \mathbf{b}}{\mathbf{c} \cdot \mathbf{d}}$.

(b) $\mathbf{a} \cdot \mathbf{b} \cdot \mathbf{c}$,

---

[1] The point of this exercise is to emphasize basic algebraic properties of vectors. The dot product of two vectors is a scalar, the cross-product of two vectors is a vector, and dividing by a vector is not defined.

(c) $\frac{\mathbf{a} \cdot \mathbf{b} \, \mathbf{d}}{\mathbf{b} \cdot \mathbf{c}}$.

(d) $\frac{\mathbf{a} \cdot \mathbf{d} \, \mathbf{b} \cdot \mathbf{c}}{\mathbf{d}}$.

(e) $|\mathbf{a} \cdot \mathbf{b} \, \mathbf{d}|$,

(f) $\mathbf{a} |\mathbf{c} - \mathbf{d}|$,

(g) $\mathbf{a} + \mathbf{b} \cdot \mathbf{c}$,

(h) $\mathbf{a} \times \mathbf{b}$,

(i) $\mathbf{a} \times \mathbf{b} \times \mathbf{c}$,

(j) $\mathbf{a} \times \mathbf{b} + \mathbf{a} \cdot \mathbf{b}$,

(k) $\mathbf{a} \times \mathbf{b} \cdot \mathbf{d}$,

(l) $\mathbf{a} \times (\mathbf{b} \cdot \mathbf{c})$.

## Solution.

(a) $\mathbf{a} \cdot \mathbf{b} = 1$, $\mathbf{c} \cdot \mathbf{d} = 1$, hence $\frac{\mathbf{a} \cdot \mathbf{b}}{\mathbf{c} \cdot \mathbf{d}} = 1$.

(b) This expression is not mathematically valid.

(c) The vector operations are all fine. However, the denominator is zero since $\mathbf{b} \cdot \mathbf{c} = 0$.

(d) Dividing by a vector is not defined.

(e) $(\mathbf{a} \cdot \mathbf{b}) \mathbf{d} = \mathbf{i} + \mathbf{j}$ and $|\mathbf{i} + \mathbf{j}| = \sqrt{2}$.

(f) $|\mathbf{c} - \mathbf{d}| = |-\mathbf{i}| = 1$, therefore $\mathbf{a} |\mathbf{c} - \mathbf{d}| = \mathbf{a} = \mathbf{i} - \mathbf{j}$,

(g) This would be much clearer if there were parentheses in appropriate places in the expression. The usual rules of algebra require that multiplications are performed first, then addition. The same holds in vector algebra. In this case we compute the dot product of $\mathbf{b}$ and $\mathbf{c}$ first (which is a scalar), then add it to $\mathbf{a}$. However, adding a scalar to a vector is not defined.

(h) $\mathbf{a} \times \mathbf{b} = (\mathbf{i} - \mathbf{j}) \times \mathbf{i} = \mathbf{k}$.

(i) This is ambiguous since $(\mathbf{a} \times \mathbf{b}) \times \mathbf{c}$, is not equal to $\mathbf{a} \times (\mathbf{b} \times \mathbf{c})$, and therefore you do not know which cross-product to compute first.

(j) This expression describes the addition of a vector and a scalar, which is not a mathematically valid operation.

(k) This expression might appear ambiguous since it is not clear whether to evaluate the cross-product first or the dot product, and evaluating the dot product first would give nonsense. However, in the literature there is a convention for such expressions involving cross-products and dot products. It is $\mathbf{a} \times \mathbf{b} \cdot \mathbf{d} \equiv (\mathbf{a} \times \mathbf{b}) \cdot \mathbf{d}$. In this case we have $(\mathbf{a} \times \mathbf{b}) \cdot \mathbf{d} = \mathbf{k} \cdot (\mathbf{i} \cdot \mathbf{j}) = 0$.

(l) This expression is not mathematically valid since you cannot compute the cross-product of a vector with a scalar.

**Problem 5.** Consider the vectors

$$\mathbf{A} = A_1\mathbf{i} + A_2\mathbf{j} + A_3\mathbf{k}, \quad \mathbf{B} = B_1\mathbf{i} + B_2\mathbf{j} + B_3\mathbf{k},$$

both emanating from the origin of a cartesian coordinate system. Let the tip of the vector $\mathbf{A}$ be denoted by $P$ and the tip of the vector $\mathbf{B}$ be denoted by $Q$. In other words, with respect to the cartesian coordinate system, $P$ is located at the point $(A_1, A_2, A_3)$ and $Q$ is located at the point $(B_1, B_2, B_3)$. Determine the vector starting at $P$ and ending at $Q$, and compute its magnitude. See figure below.

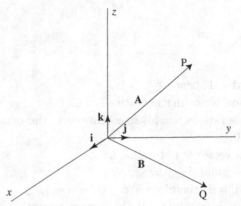

**Solution.** Let $\mathbf{PQ}$ denote the vector starting at the point $P$ and ending at the point $Q$.

$$\mathbf{A} + \mathbf{PQ} = \mathbf{B}.$$

Then

$$\mathbf{PQ} = \mathbf{B} - \mathbf{A},$$
$$= (B_1 - A_1)\mathbf{i} + (B_2 - A_2)\mathbf{j} + (B_3 - A_3)\mathbf{k},$$

and

$$|\mathbf{PQ}| = \sqrt{(B_1 - A_1)^2 + (B_2 - A_2)^2 + (B_3 - A_3)^2},$$

**Problem 6.** Let $\mathbf{A}(t) = A_1(t)\mathbf{i} + A_2(t)\mathbf{j} + A_3(t)\mathbf{k}$, $\mathbf{B}(t) = B_1(t)\mathbf{i} + B_2(t)\mathbf{j} + B_3(t)\mathbf{k}$, and $a(t)$ denote a scalar valued function. Prove the following properties for different types of products.

(a) $\frac{d}{dt}\left(a(t)\mathbf{A}(t)\right) = a(t)\frac{d\mathbf{A}}{dt}(t) + \frac{da}{dt}(t)\mathbf{A}(t),$

(b) $\frac{d}{dt}\left(\mathbf{A}(t) \cdot \mathbf{B}(t)\right) = \mathbf{A}(t) \cdot \frac{d\mathbf{B}}{dt}(t) + \frac{d\mathbf{A}}{dt}(t) \cdot \mathbf{B}(t),$

(c) $\frac{d}{dt}\left(\mathbf{A}(t) \times \mathbf{B}(t)\right) = \mathbf{A}(t) \times \frac{d\mathbf{B}}{dt}(t) + \frac{d\mathbf{A}}{dt}(t) \times \mathbf{B}(t).$

**Solution.** We will omit denoting the explicit dependence of the functions on $t$ for the sake of a less cumbersome notation.

(a) $a\mathbf{A} = aA_1\mathbf{i} + aA_2\mathbf{j} + aA_3\mathbf{k}$, then

$$\frac{d}{dt}(a\mathbf{A}) = \left(\frac{da}{dt}A_1 + a\frac{dA_1}{dt}\right)\mathbf{i} + \left(\frac{da}{dt}A_2 + a\frac{dA_2}{dt}\right)\mathbf{j}$$

$$+ \left(\frac{da}{dt}A_3 + a\frac{dA_3}{dt}\right)\mathbf{k},$$

$$= \frac{da}{dt}A_1\mathbf{i} + \frac{da}{dt}A_2\mathbf{j} + \frac{da}{dt}A_3\mathbf{k} + a\frac{dA_1}{dt}\mathbf{i} + a\frac{dA_2}{dt}\mathbf{j} + a\frac{dA_3}{dt}\mathbf{k},$$

$$= \frac{da}{dt}(A_1\mathbf{i} + A_2\mathbf{j} + A_3\mathbf{k}) + a\left(\frac{dA_1}{dt}\mathbf{i} + \frac{dA_2}{dt}\mathbf{j} + \frac{dA_3}{dt}\mathbf{k}\right),$$

$$- \frac{da}{dt}\mathbf{A} + u\frac{d\mathbf{A}}{dt}.$$

(b) $\mathbf{A} \cdot \mathbf{B} = A_1B_1 + A_2B_2 + A_3B_3$, then

$$\frac{d}{dt}(\mathbf{A}\cdot\mathbf{B}) = \left(\frac{dA_1}{dt}B_1 + A_1\frac{dB_1}{dt}\right) + \left(\frac{dA_2}{dt}B_2 + A_2\frac{dB_2}{dt}\right)$$

$$+ \left(\frac{dA_3}{dt}B_3 + A_3\frac{dB_3}{dt}\right),$$

$$= \frac{dA_1}{dt}B_1 + \frac{dA_2}{dt}B_2 + \frac{dA_3}{dt}B_3$$

$$+ A_1\frac{dB_1}{dt} + A_2\frac{dB_2}{dt} + A_3\frac{dB_3}{dt},$$

$$= \frac{d\mathbf{A}}{dt}\cdot\mathbf{B} + \mathbf{A}\cdot\frac{d\mathbf{B}}{dt}.$$

(c) $\mathbf{A} \times \mathbf{B} = (A_2B_3 - A_3B_2)\mathbf{i} + (A_3B_1 - A_1B_3)\mathbf{j} + (A_1B_2 - A_2B_1)\mathbf{k}$, then

$$\frac{d}{dt}(\mathbf{A}\times\mathbf{B}) = \left(\frac{dA_2}{dt}B_3 + A_2\frac{dB_3}{dt} - \frac{dA_3}{dt}B_2 - A_3\frac{dB_2}{dt}\right)\mathbf{i}$$

$$+ \left(\frac{dA_3}{dt}B_1 + A_3\frac{dB_1}{dt} - \frac{dA_1}{dt}B_3 - A_1\frac{dB_3}{dt}\right)\mathbf{j}$$

$$+ \left(\frac{dA_1}{dt}B_2 + A_1\frac{dB_2}{dt} - \frac{dA_2}{dt}B_1 - A_2\frac{dB_1}{dt}\right)\mathbf{k},$$

$$= \left(\frac{dA_2}{dt}B_3 - \frac{dA_3}{dt}B_2\right)\mathbf{i} + \left(\frac{dA_3}{dt}B_1 - \frac{dA_1}{dt}B_3\right)\mathbf{j}$$

$$+ \left( \frac{dA_1}{dt} B_2 - \frac{dA_2}{dt} B_1 \right) \mathbf{k}$$

$$+ \left( A_2 \frac{dB_3}{dt} - A_3 \frac{dB_2}{dt} \right) \mathbf{i} + \left( A_3 \frac{dB_1}{dt} - A_1 \frac{dB_3}{dt} \right) \mathbf{j}$$

$$+ \left( A_1 \frac{dB_2}{dt} - A_2 \frac{dB_1}{dt} \right) \mathbf{k},$$

$$= \frac{d\mathbf{A}}{dt} \times \mathbf{B} + \mathbf{A} \times \frac{d\mathbf{B}}{dt}.$$

**Problem 7.** If $\mathbf{A}(t) = 4(t-1)\mathbf{i} - (2t+3)\mathbf{j} + 6t^2\mathbf{k}$, then compute[2]:

(a) $\int_2^3 \mathbf{A}(t)dt$,

(b) $\int_1^2 (t\mathbf{i} - 2\mathbf{k}) \cdot \mathbf{A}(t)dt$.

**Solution.**

(a)

$$\int_2^3 4(t-1)dt\mathbf{i} - (2t+3)dt\mathbf{j} + 6t^2dt\mathbf{k},$$

$$= (2t^2 - 4t) \Big|_2^3 \mathbf{i} - (t^2 + 3t) \Big|_2^3 dt\mathbf{j} + 2t^3 \Big|_2^3 \mathbf{k},$$

$$= 6\mathbf{i} - 8\mathbf{j} + 38\mathbf{k}.$$

(b)

$$\int_1^2 (t\mathbf{i} - 2\mathbf{k}) \cdot \mathbf{A}(t)dt = \int_1^2 \left( 4(t^2 - t) - 12t^2 \right) dt$$

$$= -\int_1^2 (8t^2 + 4t) \, dt = -\frac{74}{3}.$$

**Problem 8.** Suppose a particle moves along a space curve defined by:
$$x(t) = e^{-t}\cos t, \ y(t) = e^{-t}\sin t, \ z(t) = e^{-t}.$$
Find the magnitude of the velocity and acceleration at any time $t$.

---

[2]This problem can be a bit tricky. It is important to understand the nature of each integrand.

**Solution.** We have $x(t) = e^{-t}\cos t$, $y(t) = e^{-t}\sin t$, $z(t) = e^{-t}$, then:

$$\dot{x} = -e^{-t}\cos t - e^{-t}\sin t, \quad \dot{y} = -e^{-t}\sin t + e^{-t}\cos t, \quad \dot{z} = -e^{-t},$$

$$\ddot{x} = e^{-t}\cos t + e^{-t}\sin t + e^{-t}\sin t - e^{-t}\cos t = 2e^{-t}\sin t,$$

$$\ddot{y} = e^{-t}\sin t - e^{-t}\cos t - e^{-t}\cos t - e^{-t}\sin t = -2e^{-t}\cos t,$$

$$\ddot{z} = e^{-t}.$$

Then we have:

$$|\mathbf{v}| = \sqrt{\dot{x}^2 + \dot{y}^2 + \dot{z}^2} = e^{-t}\sqrt{(\cos t + \sin t)^2 + (\cos t - \sin t)^2 + 1} = \sqrt{3}e^{-t}.$$

Similarly,

$$|\mathbf{a}| = \sqrt{\ddot{x}^2 + \ddot{y}^2 + \ddot{z}^2} = e^{-t}\sqrt{(2\sin t)^2 + (-2\cos t)^2 + 1} = \sqrt{5}e^{-t}.$$

**Problem 9.** The position vector of a particle at any time $t$ is given by:

$$\mathbf{r} = a\cos\omega t\mathbf{i} + a\sin\omega t\mathbf{j} + bt^2\mathbf{k},$$

where $a$ and $b$ are scalars. Show that the speed of the particle increases with time, but the magnitude of the acceleration is constant. Describe the motion of the particle geometrically.

**Solution.** We have $\mathbf{r} = a\cos\omega t\mathbf{i} + a\sin\omega t\mathbf{j} + bt^2\mathbf{k}$, then:

$$\mathbf{v} = -a\omega\sin\omega t\mathbf{i} + a\omega\cos\omega t\mathbf{j} + 2bt\mathbf{k},$$

$$\mathbf{a} = -a\omega^2\cos\omega t\mathbf{i} - a\omega^2\sin\omega t\mathbf{j} + 2b\mathbf{k}.$$

and

$$|\mathbf{v}| = \sqrt{a^2\omega^2(\sin^2\omega t + \cos^2\omega t) + (2bt)^2} = \sqrt{a^2\omega^2 + (2bt)^2},$$

$$|\mathbf{a}| = \sqrt{a^2\omega^4(\cos^2\omega t + \sin^2\omega t) + (2b)^2} = \sqrt{a^2\omega^4 + (2b)^2}.$$

The particle follows the path of a helix in three dimensions.

# Chapter 3

# Solutions for Problem Set 3

**Problem 1.** Let $C$ be a space curve and let the position of any point on the curve be given by the following vector:

$$\mathbf{r} = 3\cos 2t\mathbf{i} + 3\sin 2t\mathbf{j} + (8t - 4)\mathbf{k}.$$

(a) Find a unit tangent vector $\mathbf{T}$ to the curve (for any point on the curve).

(b) If $\mathbf{r}$ is the position vector of a particle moving on $C$ at any time $t$, verify in this case that $\mathbf{v} = v\mathbf{T}$.

(c) Compute the curvature at any point on the curve.

(d) Compute the radius of curvature at any point on the curve.

(e) Compute the unit principal normal $\mathbf{N}$ at any point on the curve.

**Solution.** We have

$$\mathbf{r} = 3\cos 2t\mathbf{i} + 3\sin 2t\mathbf{j} + (8t - 4)\mathbf{k},$$

then

$$\mathbf{v} = -6\sin 2t\mathbf{i} + 6\cos 2t\mathbf{j} + 8\mathbf{k},$$

and

$$\frac{ds}{dt} = v = |\mathbf{v}| = \sqrt{\mathbf{v}\cdot\mathbf{v}} = \sqrt{36 + 64} = 10.$$

(a)

$$\mathbf{T} = \frac{\mathbf{v}}{v} = -\frac{3}{5}\sin 2t\mathbf{i} + \frac{3}{5}\cos 2t\mathbf{j} + \frac{4}{5}\mathbf{k}.$$

(b) This should be clear.

(c) $\kappa = \left|\frac{d\mathbf{T}}{ds}\right|$. We have:

$$\frac{d\mathbf{T}}{ds} = \frac{d\mathbf{T}}{dt}\frac{dt}{ds} = \left(-\frac{6}{5}\cos 2t\mathbf{i} - \frac{6}{5}\sin 2t\mathbf{j}\right)\frac{1}{10}.$$

Therefore

$$\kappa = \frac{3}{25}.$$

(d) $R = \frac{1}{\kappa} = \frac{25}{3}$.

(e) $\mathbf{N} = R\frac{d\mathbf{T}}{ds} = \frac{25}{3}\left(-\frac{6}{5}\cos 2t\mathbf{i} - \frac{6}{5}\sin 2t\mathbf{j}\right)\frac{1}{10}$.

**Problem 2.** A particle moves so that its position vector is given by:

$$\mathbf{r} = \cos\omega t\mathbf{i} + \sin\omega t\mathbf{j},$$

where $\omega$ is a constant. Prove the following:

(a) the velocity $\mathbf{v}$ of the particle is perpendicular to $\mathbf{r}$,
(b) the acceleration $\mathbf{a}$ is directed toward the origin and has magnitude proportional to the distance from the origin,[1]
(c) $\mathbf{r} \times \mathbf{v}$ is a constant vector.[2]

**Solution.** We have:

$$\mathbf{r} = \cos\omega t\mathbf{i} + \sin\omega t\mathbf{j},$$

from which it follows that:

$$\mathbf{v} = -\omega\sin\omega t\mathbf{i} + \omega\cos\omega t\mathbf{j},$$

and

$$\mathbf{a} = -\omega^2\cos\omega t\mathbf{i} - \omega^2\sin\omega t\mathbf{j}.$$

(a) This is a trivial calculation.
(b) $\mathbf{a} = -\omega^2\mathbf{r}$
(c) $\mathbf{r} \times \mathbf{v} = \omega\mathbf{k}$.

---

[1] In terms of the vector nature of $\mathbf{a}$, it is important to understand what "directed towards the origin" means, and what "proportional to the distance from the origin" means.

[2] What does "constant vector" mean? Hint: a vector has *length* and *direction*.

**Problem 3.** Let $\mathbf{r}(t)$ denote a position vector, and consider the function:

$$T = \frac{1}{2} m \dot{\mathbf{r}} \cdot \dot{\mathbf{r}},$$

where $m$ is a constant. Compute $\frac{dT}{dt}$.

**Solution.** $m\ddot{\mathbf{r}} \cdot \dot{\mathbf{r}}$.

**Problem 4.** Let $V(\mathbf{r})$ be a scalar valued function of the position vector $\mathbf{r}(t)$. Compute $\frac{dV}{dt}$, and express it as the dot product of two vectors.

**Solution.** $\nabla V(\mathbf{r}) \cdot \dot{\mathbf{r}}$.

**Problem 5.** Consider the space curve defined by the following position vector:

$$\mathbf{r}(t) = \cos t\,\mathbf{i} + \sin t\,\mathbf{j} + t\mathbf{k},$$

and the scalar valued function:

$$V(x, y, z) = \frac{1}{2}\left(x^2 + y^2 + z^2\right).$$

Evaluate the function on the space curve, and then compute its derivative with respect to $t$.

**Solution.** Use the previous problem:

$$\frac{dV}{dt} = (x\mathbf{i} + y\mathbf{j} + z\mathbf{k}) \cdot (-\sin t\,\mathbf{i} + \cos t\,\mathbf{j} + \mathbf{k}),$$

$$= (\cos t\,\mathbf{i} + \sin t\,\mathbf{j} + t\mathbf{k}) \cdot (-\sin t\,\mathbf{i} + \cos t\,\mathbf{j} + \mathbf{k}) = t.$$

**Problem 6.** Consider the space curve defined by the following position vector:

$$\mathbf{r}(t) = \cos t\,\mathbf{i} + \sin t\,\mathbf{j} + t\mathbf{k}.$$

Compute the length of a piece of the curve from the point $(1, 0, 0)$ to the point $\left(0, 1, \frac{\pi}{2}\right)$.

**Solution.** First, you should have verified that the two points are on the curve. Then recall the definition of arclength, $s$:

$$\frac{ds}{dt} \equiv \sqrt{\left(\frac{dx}{dt}(t)\right)^2 + \left(\frac{dy}{dt}(t)\right)^2 + \left(\frac{dz}{dt}(t)\right)^2}.$$

So for this problem we have:

$$\text{length} = \int_0^{\frac{\pi}{2}} \sqrt{2}\,dt = \frac{\pi}{\sqrt{2}}.$$

It is crucial that you understand the reason for the choice of the limits in the integral.

**Problem 7.** Parametrize the curve in the previous problem in terms of arclength, rather than $t$.

**Solution.** Use the indefinite integral from the previous problem to compute arclength as a function of $t$:

$$s = \int_0^s ds = \int_0^t \sqrt{2}\,dt = \sqrt{2}t.$$

(Why were the limits on the integrals chosen as above?) Then we have:

$$\mathbf{r}(t) = \cos t\mathbf{i} + \sin t\mathbf{j} + t\mathbf{k} = \cos\frac{1}{\sqrt{2}}s\mathbf{i} + \sin\frac{1}{\sqrt{2}}s\mathbf{j} + \frac{1}{\sqrt{2}}s\mathbf{k} = \mathbf{r}(t(s)) = \mathbf{r}(s).$$

(Make sure you understand what is meant by the four equality signs in the expression above.)

**Problem 8.** For the example in the previous two exercises show explicitly that:

$$\frac{d\mathbf{r}}{dt} = \frac{d\mathbf{r}}{ds}\frac{ds}{dt}.$$

**Solution.**

$$\frac{d\mathbf{r}}{dt} = -\sin t\mathbf{i} + \cos t\mathbf{j} + \mathbf{k}.$$

$$\frac{d\mathbf{r}}{ds} = \frac{1}{\sqrt{2}}\left(-\sin\frac{1}{\sqrt{2}}s\mathbf{i} + \cos\frac{1}{\sqrt{2}}s\mathbf{j} + \mathbf{k}\right).$$

$$\frac{ds}{dt} = \sqrt{2}.$$

# Chapter 4

# Solutions for Problem Set 4

**Problem 1.** If $\mathbf{A} = (3x^2 - 6yz)\mathbf{i} + (2y + 3xz)\mathbf{j} + (1 - 4xyz^2)\mathbf{k}$ evaluate $\int_C \mathbf{A} \cdot d\mathbf{r}$ from $(0,0,0)$ to $(1,1,1)$ along the following paths $C$:

(a) $x = t$, $y = t^2$, $z = t^3$,

(b) the straight line joining $(0,0,0)$ to $(1,1,1)$.

**Solution.** First,

$$\int_C \mathbf{A} \cdot d\mathbf{r} = \int_C \left((3x^2 - 6yz)\mathbf{i} + (2y + 3xz)\mathbf{j} + (1 - 4xyz^2)\mathbf{k}\right)$$
$$\cdot (dx\mathbf{i} + dy\mathbf{j} + dz\mathbf{k}),$$
$$= \int_C (3x^2 - 6yz)dx + (2y + 3xz)dy + (1 - 4xyz^2)dz.$$

(a) If $x = t$, $y = t^2$, $z = t^3$, then the points $(0,0,0)$ and $(1,1,1)$ correspond to $t = 0$ and $t = 1$, respectively. Then we have

$$\int_C \mathbf{A} \cdot d\mathbf{r} = \int_{t=0}^{t=1} (3t^2 - 6t^5)dt + (2t^2 + 3t^4)d(t^2) + (1 - 4t^9)d(t^3),$$
$$= \int_{t=0}^{t=1} (3t^2 - 6t^5)dt + (4t^3 + 6t^5)dt + 3(t^2 - 4t^{11})dt,$$
$$= \left(t^3 - t^6 + t^4 + t^5 + t^3 - t^{12}\right)\Big|_0^1 = 2.$$

(b) Along the straight line joining $(0,0,0)$ to $(1,1,1)$ we have $x = t, y = t, z = t$. Then since $dx = dy = dz = dt$, we have:

$$\int_C \mathbf{A} \cdot d\mathbf{r} = \int_C (3x^2 - 6yz)dx + (2y + 3xz)dy + (1 - 4xyz^2)dz,$$

$$= \int_0^1 (3t^2 - 6t^2)dt + (2t + 3t^2)dt + (1 - 4t^4)dt,$$

$$= \left( -t^3 + t^2 + t^3 + t - \frac{4}{5}t^5 \right) \Big|_0^1 = \frac{6}{5}.$$

**Problem 2.** Consider the following scalar valued function of $(x, y, z)$:

$$\phi(x, y, z) = x^3 + zy + xy + \sin xy + \cos \frac{x^2}{z}.$$

Compute $\frac{\partial \phi}{\partial x}$, $\frac{\partial \phi}{\partial y}$, and $\frac{\partial \phi}{\partial z}$.

**Solution.**

$$\frac{\partial \phi}{\partial x} = 3x^2 + y + y \cos xy - \frac{2x}{z} \sin \frac{x^2}{z}.$$

$$\frac{\partial \phi}{\partial y} = z + x + x \cos xy.$$

$$\frac{\partial \phi}{\partial z} = y + \frac{x^2}{z^2} \sin \frac{x^2}{z}.$$

**Problem 3.** Suppose

$$\mathbf{A} = (2xy + z^3)\mathbf{i} + (x^2 + 2y)\mathbf{j} + (3xz^2 - 2)\mathbf{k}.$$

Show that the line integral of $\mathbf{A}$ between two points is independent of the path taken between the two points.

**Solution.** Show that $\nabla \times \mathbf{A} = 0$.

$$\frac{\partial A_1}{\partial x} = 2y, \quad \frac{\partial A_1}{\partial y} = 2x, \quad \frac{\partial A_1}{\partial z} = 3z^2.$$

$$\frac{\partial A_2}{\partial x} = 2x, \quad \frac{\partial A_2}{\partial y} = 2, \quad \frac{\partial A_2}{\partial z} = 0.$$

$$\frac{\partial A_3}{\partial x} = 3z^2, \quad \frac{\partial A_3}{\partial y} = 0, \quad \frac{\partial A_3}{\partial z} = 6xz.$$

Now it is easy to verify that:

$$\frac{\partial A_3}{\partial y} - \frac{\partial A_2}{\partial z} = \frac{\partial A_1}{\partial z} - \frac{\partial A_3}{\partial x} = \frac{\partial A_2}{\partial x} - \frac{\partial A_1}{\partial y} = 0.$$

**Problem 4.** Consider the following vector fields:

(a) $\mathbf{A}(x, y, z) = \cos x \sin y \sin z \mathbf{i} + \sin x \cos y \sin z \mathbf{j} + \sin x \sin y \cos z \mathbf{k}$,
(b) $\mathbf{A}(x, y, z) = yz\mathbf{i} + xz\mathbf{j} + xy\mathbf{k}$,
(c) $\mathbf{A}(x, y, z) = z\mathbf{k}$,

Compute the line integral of $\mathbf{A}$ along any path of your choice connecting $(0, 0, 0)$ to $(1, 1, 1)$.

**Solution.** First, note that each vector field is the gradient of a scalar valued function, $\mathbf{A} = \nabla V$. Therefore, the line integral of the vector along a path between two points is the difference of the scalar valued function evaluated at the two points.

(a) $V = \sin x \sin y \sin z$. $V(1, 1, 1) - V(0, 0, 0) = (\sin 1)^3$.
(b) $V = xyz$. $V(1, 1, 1) - V(0, 0, 0) = 1$.
(c) $V = \frac{z^2}{2}$. $V(1, 1, 1) - V(0, 0, 0) = \frac{1}{2}$.

# Chapter 5

# Solutions for Problem Set 5

**Problem 1.** Consider a projectile that is launched with a velocity of magnitude $v_0$ at an angle of $\alpha$ with respect to the horizontal. Assume that the only force acting on the projectile is gravity, which is further assumed to be constant, and acting vertically downward (and we may take the vertical coordinate to be $z$).

(a) With no loss of generality the motion could be assumed to be in the $y - z$ plane. Explain why.[1]

(b) Give an argument that the $y$ component of velocity is constant in time.[2]

**Solution.** The answer to both parts of this question involve an understanding of Newton's first law: velocity does not change unless there is a force, and the velocity changes in the direction of the force.

(a) The initial velocity vector and the vector defining the gravitational force define a plane, which we label as the $y - z$ plane. Motion must occur in this plane since there is no force acting "out of the plane."

(b) There is no force in the $y$ direction. Therefore, the initial velocity in the $y$ direction never changes.

**Problem 2.** A particle of mass $m$ moves along a straight line (which, without loss of generality (why?) we may consider to be the $x$-axis) under the influence of a constant force $F$, see figure below.

---

[1] Hint: What does Newton's first law have to say about the situation?
[2] Hint: Again, consider Newton's first law.

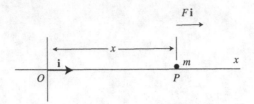

Suppose that the particle starts at $x = 0$ at $t = 0$ with a velocity $v_0\mathbf{i}$ ($v_0 > 0$). Find:

(a) the speed,
(b) the velocity as a function of time,
(c) the distance traveled after time $t$,
(d) the speed as a function of position $(x)$. (Hint: use the previous two results and eliminate time between them.)

**Solution.** The starting point is Newton's equations, which are given by:

$$m\frac{d^2x}{dt^2}\mathbf{i} = F\mathbf{i}, \quad x(0) = 0, \, \dot{x}(0) = v_0.$$

We integrate once (with respect to time) to get speed and velocity:

$$\int_0^t \frac{d}{d\tau}\left(\frac{dx}{d\tau}(\tau)\right)d\tau\mathbf{i} = \frac{F}{m}\int_0^t d\tau\mathbf{i},$$

or

$$\frac{dx}{dt}(t)\mathbf{i} = \left(v_0 + \frac{F}{m}t\right)\mathbf{i},$$

which gives the velocity as a function of time. The speed is the magnitude of velocity:

$$\frac{dx}{dt}(t) = v_0 + \frac{F}{m}t. \tag{5.1}$$

To get distance we integrate the expression for velocity (with respect to time):

$$\int_0^t \frac{dx}{d\tau}(\tau)d\tau = \int_0^t \left(v_0 + \frac{F}{m}\tau\right)d\tau,$$

or

$$x(t) = v_0 t + \frac{F}{2m}t^2. \tag{5.2}$$

Finally, we solve for speed as a function of position. Start with (5.2). This is a quadratic equation for $t$ that we can solve for $t$:

$$t = -\frac{mv_0}{F} \pm \frac{m}{F}\sqrt{v_0^2 + \frac{2Fx(t)}{m}}.$$

There are two choices of sign here. Which one do we take? Now $t$ is positive (we start from $t = 0$ and $t$ increases). The constants $m$, $F$, and $v_0$ are all positive, which implies that $x(t)$ is positive (look at (5.2)). Therefore for $t$ positive we must have:

$$t = -\frac{mv_0}{F} + \frac{m}{F}\sqrt{v_0^2 + \frac{2Fx(t)}{m}}.$$

Substituting this into (5.1) (and writing $\frac{dx}{dt}(t) = v(t)$) gives:

$$v(t) = \sqrt{v_0^2 + \frac{2Fx(t)}{m}}, \tag{5.3}$$

or

$$(v(t))^2 = v_0^2 + \frac{2Fx(t)}{m}.$$

**Problem 3.** An object of mass $m$ is thrown vertically upward from the Earth's surface with an initial velocity $v_0\mathbf{k}$ ($v_0 > 0$). We assume that the only force acting on the object is gravity, see figure below.

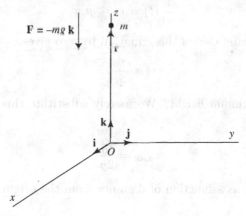

Find:

(a) the position at any time,
(b) the time taken to reach the highest point,
(c) the maximum height reached,
(d) the speed as a function of its distance from the origin.

**Solution.** We denote the position vector of the object by $\mathbf{r} = z\mathbf{k}$. The Newton's equations become:

$$m\frac{d^2z}{dt^2}\mathbf{k} = -mg\mathbf{k}, \quad z(0) = 0, \, \dot{z}(0) = v_0 > 0,$$

or,

$$\frac{d^2 z}{dt^2}\mathbf{k} = -g\mathbf{k}, \quad z(0) = 0, \ \dot{z}(0) = v_0 > 0,$$

Now these equations are identical to those of the previous problem with

$$\frac{F}{m} = -g.$$

So, using (5.2), we have

$$z(t) = v_0 t - \frac{g}{2}t^2. \tag{5.4}$$

Next we need to compute the time taken to reach the highest point. We must ask ourselves, "what characterizes the highest point"? The object goes up, stops "instantaneously", and falls back down. So the highest point is reached at the time when the speed vanishes.

Using (5.1), we have:

$$\frac{dz}{dt}(t) = v_0 - gt. \tag{5.5}$$

Setting the left-hand side of this equation to zero gives:

$$t = \frac{v_0}{g}.$$

What is the maximum height? We merely substitute this time into (5.4) to get:

$$z_{\max} = \frac{v_0^2}{2g}.$$

To get the speed as a function of distance from the origin we use (5.3) to obtain:

$$v(t) = \sqrt{v_0^2 - 2gz(t)}. \tag{5.6}$$

**Problem 4.** Now we consider a variation of the previous problem. Suppose we drop an object of mass $m$ from a height $h$ at $t = 0$ (from this statement it should be clear that the speed of the object at $t = 0$ is 0). In addition to the force of gravity, suppose there is a force (due to air resistance) directed in the positive $z$ direction that is proportional to the instantaneous speed, i.e. the force has the form $\beta v\mathbf{k}$, where $\beta$ is a positive scalar, see figure below.

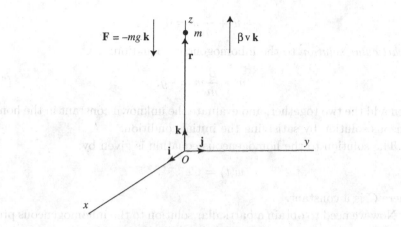

Find:

(a) the speed as a function of time,
(b) the distance traveled as a function of time,
(c) the acceleration at any time $t > 0$.

Show that there is a limiting speed, i.e. as time increases, the speed approaches a limit.

**Solution.** First, we write down Newton's equations:

$$m\frac{d^2z}{dt^2}\mathbf{k} = -mg\mathbf{k} - \beta\frac{dz}{dt}\mathbf{k}, \quad z(0) = h, \ \dot{z}(0) = 0,$$

or

$$\dot{w} + \frac{\beta}{m}w = -g, \quad w(0) = 0, \tag{5.7}$$

where $w \equiv \frac{dz}{dt}$. As discussed in class, this is a *linear, inhomogeneous first order equation for* $w$. We solve for $w$, then integrate the result to get the height.

To find the general solution of (5.7), we find a solution to the homogeneous equation:

$$\dot{w} + \frac{\beta}{m}w = 0,$$

a *particular solution* to the inhomogeneous equation:

$$\dot{w} + \frac{\beta}{m}w = -g. \tag{5.8}$$

then add the two together, and evaluate the unknown constant in the homogeneous solution by satisfying the initial condition.

The solution to the homogeneous equation is given by:

$$w(t) = Ce^{-\frac{\beta}{m}t},$$

where $C$ is a constant.

Now we need to obtain a particular solution to the inhomogeneous problem. There is a general method for this. But this problem has a particular structure that makes it simple. Look at the right-hand side of (5.8). It is a constant. The derivative of a constant is zero. Now look at the left-hand side of (5.8). It has a term that is a derivative of $w$, plus a constant time $w$. Hence, it follows that we can find a solution of the form $w = $ constant. In this case:

$$w_p = -\frac{mg}{\beta}.$$

Then the general solution is:

$$w(t) = Ce^{-\frac{\beta}{m}t} - \frac{mg}{\beta}.$$

Now $w(0) = 0$, so we have:

$$w(0) = C - \frac{mg}{\beta} = 0,$$

or

$$C = \frac{mg}{\beta},$$

and therefore:

$$w(t) = \frac{mg}{\beta}e^{-\frac{\beta}{m}t} - \frac{mg}{\beta}.$$

or

$$\frac{dz}{dt} = w(t) = \frac{mg}{\beta}\left(e^{-\frac{\beta}{m}t} - 1\right).$$

This gives the speed as a function of time. We easily see that there is a limiting speed since:

$$\lim_{t\to\infty} w(t) = \lim_{t\to\infty} \frac{mg}{\beta}\left(e^{-\frac{\beta}{m}t} - 1\right) = -\frac{mg}{\beta}.$$

We could quickly get the acceleration as a function of time by differentiating the expression for the velocity as a function of time:

$$\ddot{z} = -ge^{-\frac{\beta}{m}t}.$$

To obtain the position as a function of time we integrate the expression for the velocity as a function of time:

$$\int_0^t \frac{dz}{d\tau}(\tau)d\tau = \int_0^t \left(\frac{mg}{\beta}\left(e^{-\frac{\beta}{m}\tau} - 1\right)\right)d\tau,$$

which gives:

$$z(t) = h - \frac{mg}{\beta}t - \frac{m^2g}{\beta^2}\left(e^{-\frac{\beta}{m}t} - 1\right).$$

**Problem 5.** Recall the definition of linear ODE given in this chapter:

$$m\frac{d^2s}{dt^2} = (a_0 + a_1(t))s + (b_0 + b_1(t))\dot{s} + c_0 + c_1(t), \qquad (5.19)$$

where $a_0, b_0, c_0$ are constants, and $a_1(t), b_1(t), c_1(t)$ are functions of $t$.
    First, consider the situation where $c_0 = c_1(t) = 0$, i.e.

$$m\frac{d^2s}{dt^2} = (a_0 + a_1(t))s + (b_0 + b_1(t))\dot{s}. \qquad (5.20)$$

In this case the linear ODE is said to be homogeneous.

(a) Suppose $s_1(t)$ is a solution of (5.20), and let $k_1$ denote a constant (real number). Prove that $k_1 s_1(t)$ is also a solution of (5.20).
(b) Suppose $s_1(t)$ and $s_2(t)$ are solutions of (5.20), and let $k_1$ and $k_2$ denote constants (real numbers). Prove that $k_1 s_1(t) + k_2 s_2(t)$ is also a solution of (5.20). This is the *superposition principle* for linear homogeneous ODE's.
(c) Do these two results hold for (5.19)?

(d) Now let us consider a physical application of the results in (a) and
(b) applied to an ODE of the type (5.20). Consider a particle of mass
$m$ moving vertically under the influence of a (constant) gravitational
force. Suppose at $t = 0$ the particle is at height 12 (in some units that
are not important for this question) with velocity zero. We drop the
ball and its position as a function of time is $\tilde{s}(t)$ for $t > 0$.

Now suppose that we consider a different situation. Suppose at $t = 0$
the particle is at height 24 with zero velocity. What is the particle's
position as a function of time for $t > 0$?

**Solution.** Substitute the proposed solution into the ODE and see if it
indeed satisfies the ODE.

(a) We need to show that:

$$m\frac{d^2(k_1 s_1)}{dt^2} - (a_0 + a_1(t))(k_1 s_1) - (b_0 + b_1(t))\frac{d}{dt}(k_1 s_1) = 0.$$

but this is the same as:

$$k_1\left(m\frac{d^2 s_1}{dt^2} - (a_0 + a_1(t))s_1 - (b_0 + b_1(t))\dot{s}_1\right) = 0.$$

and we know that the expression in parentheses is zero since $s_1(t)$ is a
solution.

(b) We need to show that:

$$m\frac{d^2(k_1 s_1 + k_2 s_2)}{dt^2} - (a_0 + a_1(t))(k_1 s_1 + k_2 s_2))$$

$$-(b_0 + b_1(t))\frac{d}{dt}(k_1 s_1 + k_2 s_2) = 0.$$

but this is the same as:

$$k_1\left(m\frac{d^2 s_1}{dt^2} - (a_0 + a_1(t))s_1 - (b_0 + b_1(t))\dot{s}_1\right)$$

$$+k_2\left(m\frac{d^2 s_2}{dt^2} - (a_0 + a_1(t))s_2 - (b_0 + b_1(t))\dot{s}_2\right) = 0.$$

and we know that the expressions in parentheses are zero since $s_1(t)$
and $s_2(t)$ are solutions.

(c) No.

(d) What do you know? You know that $\tilde{s}(t)$ is a solution of Newton's
equations satisfying $\tilde{s}(0) = 12$ and $\frac{d\tilde{s}}{dt}(0) = 0$. What we would like
to find is a solution of Newton's equations, $\hat{s}(t)$ satisfying $\hat{s}(0) = 24$

and $\frac{d\hat{s}}{dt}(0) = 0$. It is very tempting to appeal to the linear properties of the equation by setting $\hat{s}(t) \equiv 2\tilde{s}(t)$, then $\hat{s}(0) = 2\tilde{s}(0) = 24$ and $\frac{d\hat{s}}{dt}(0) = 2\frac{d\tilde{s}}{dt}(0) = 0$. However, this is not correct since these properties only apply to linear *homogeneous* equations. So we have:

$$\tilde{s}(t) = \tilde{s}(0) - \frac{1}{2}gt^2$$

$$\hat{s}(t) = \hat{s}(0) - \frac{1}{2}gt^2.$$

The term $\frac{1}{2}gt^2$ is due to the inhomogeneous term in Newton's equations.

**Problem 6.** Are the following ordinary differential equations linear or nonlinear?

(a) $m\ddot{s} = -s + \cos t$,
(b) $m\ddot{s} = -s^2 + \cos t$,
(c) $m\ddot{s} = -s\cos t$,
(d) $m\ddot{s} = -t^2 s$,
(e) $m\ddot{s} = s \mid s^2$,

**Solution.**

(a) linear,
(b) nonlinear,
(c) linear,
(d) linear,
(e) nonlinear.

**Problem 7.** Solve the following ODE:

$$m\ddot{s} = g, \quad s(0) = s_0, \quad \dot{s}(0) = 0,$$

where $g$ is a constant.

**Solution.** In this chapter we showed that the general solution of Newton's equation in one dimension for a constant force is:

$$s(t) = s_0 + v_0(t - t_0) + \frac{F}{2m}(t - t_0)^2.$$

So for this problem we have:

$$s(t) = s_0 + \frac{g}{2m}t^2.$$

**Problem 8.** Solve the following ODE:

$$m\ddot{s} = \sin t, \quad s(0) = s_0, \ \dot{s}(0) = 0.$$

**Solution.** In this chapter we showed that the general solution of Newton's equation in one dimension for a purely time-dependent force is:

$$s(t) = s_0 + v_0(t - t_0) + \frac{1}{m} \int_{t_0}^{t} \int_{t_0}^{\tau'} F(\tau)d\tau d\tau'.$$

So for this problem we have:

$$s(t) = s_0 + \frac{1}{m}(t - \sin t).$$

Does this result make sense? The force is bounded, and its average value is zero. Yet, according to the solution for the position as a function of time, the particle moves to infinity as $t \to \infty$.

**Problem 9.** Consider the following ODE:

$$m\ddot{s} = s - s^2.$$

Find the function of $s$ and $\dot{s}$ that a solution of this ODE must satisfy.

**Solution.** From this chapter we define:

$$V(s) = -\int_{c}^{s} (s' - s'^2)ds',$$

where $c$ is a "conveniently chosen" constant. Choosing $c = 0$ we have:

$$V(s) = -\frac{s^2}{2} + \frac{s^3}{3}.$$

Then we showed that all solutions must satisfy:

$$\frac{m}{2}\dot{s}^2 - \frac{s^2}{2} + \frac{s^3}{3} = \text{constant}.$$

What do we mean by "all solutions"? Where are the initial conditions? You will see plenty of this later in the course.

**Problem 10.** Consider Newton's equations:

$$m\frac{d^2s}{dt^2} = F(s).$$

Define a new time variable, $\tau$, which is related to the "old" time $t$ by:

$$t = \sqrt{m}\tau.$$

Use the chain rule to show that with respect to the new time the ODE becomes:

$$\frac{d^2 s}{d\tau^2} = F(s),$$

i.e. the constant disappears. This is referred to as *rescaling time*.

**Solution.** With $t = \sqrt{m}\tau$ we have:

$$\frac{d}{dt} = \frac{d}{d\tau}\frac{d\tau}{dt} = \frac{1}{\sqrt{m}}\frac{d}{d\tau},$$

and

$$\frac{d^2}{dt^2} = \frac{1}{m}\frac{d^2}{d\tau^2}$$

from which the result easily follows.

**Problem 11.** Consider the following nonlinear ODE:

$$\ddot{s} = s - s^2.$$

Suppose $s_1(t)$ and $s_2(t)$ are solutions, and let $k_1$ and $k_2$ denote constants.

(a) Is $k_1 s_1(t)$ a solution?
(b) Is $k_1 s_1(t) + k_2 s_2(t)$ a solution?

**Solution.** No. This should be a trivial calculation. In general, superposition does NOT hold for nonlinear ODE's. This is ONE of the major differences. (However, there are certain exceptional situations where nonlinear ODEs can be said to obey a superposition principle.)

# Chapter 6

# Solutions for Problem Set 6

> **Key point:** There are several points to take particular care with in the problems below.
>
> - Do the distances and times that you compute make "physical sense" in the context of the stated problem? For example, are particles moving through solid obstacles? Are time intervals real and positive?
> - In some cases you will have to solve quadratic equations, which have two solutions. Make sure you determine if both solutions, one solution, or no solutions make "physical sense" in the context of the stated problem.
> - Distances are measured with respect to a chosen coordinate system. Make sure you are taking the appropriate signs for a distance with respect to the chosen coordinate system.

**Problem 1.** A particle $P$ of constant mass $m$ slides without rolling down an inclined plane of angle $\alpha$ that has a constant coefficient of friction $\mu$, see figure below.

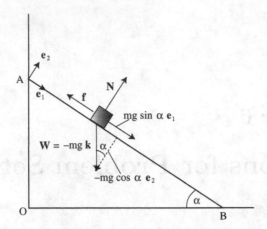

If the particle starts from rest at the top of the incline (at point $A$) find:

(a) the acceleration,
(b) the velocity,
(c) the distance traveled after time $t$.

**Solution.** We did this example in the chapter *without* friction. In this case, in addition to the forces $\mathbf{W}$ and $\mathbf{N}$ acting on $P$, there is a frictional force $\mathbf{f}$ directed up the incline (in a direction opposite to the motion) and with magnitude:

$$\mu N = \mu mg \cos \alpha, \quad \text{or}$$

$$\mathbf{f} = -\mu mg \cos \alpha \mathbf{e}_1.$$

Using this to modify Newton's equations that we derived in the chapter, you should readily see that:

$$m\frac{d^2(s\mathbf{e}_1)}{dt^2} = \mathbf{W} + \mathbf{N} + \mathbf{f},$$

$$= mg \sin \alpha \mathbf{e}_1 - \mu mg \cos \alpha \mathbf{e}_1.$$

The acceleration is given by:

$$\frac{d^2 s}{dt^2}\mathbf{e}_1 = g(\sin \alpha - \mu \cos \alpha)\mathbf{e}_1, \tag{6.1}$$

where, recall, $s$ is the distance from the top of the incline. It should be clear that we must have $\sin \alpha > \mu \cos \alpha$ or the frictional force is so great that the particle does not move at all.

Next we compute the velocity. Replacing $\frac{d^2s}{dt^2}$ by $\frac{dv}{dt}$ in (6.1), using the fact that the particle starts from rest (i.e. $v(0) = 0$), and integrating from 0 to $t$ gives the velocity:

$$v\mathbf{e}_1 = g(\sin\alpha - \mu\cos\alpha)t\mathbf{e}_1.$$

Finally, we compute the distance traveled after time $t$. Replacing $v$ with $\frac{ds}{dt}$ in the above equation, using $s(0) = 0$, and integrating from 0 to $t$ gives:

$$s = \frac{g}{2}(\sin\alpha - \mu\cos\alpha)t^2,$$

where we have dropped $\mathbf{e}_1$ since we are only interested in displacement.

**Problem 2.** Refer to figure below. An inclined plane makes an angle $\alpha$ with the horizontal.

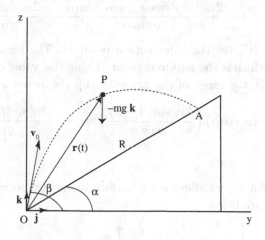

A projectile is launched from the bottom of the plane (point $O$) with speed $v_0$ in a direction making an angle $\beta$ with the horizontal. Prove that the range $R$ up the incline is given by:

$$R = \frac{2v_0^2\sin(\beta - \alpha)\cos\beta}{g\cos^2\alpha}.$$

**Solution.** In the chapter, we solved for the motion of the projectile in the absence of the incline. We found that the position vector at any time $t$ was given by:

$$\mathbf{r} = (v_0 \cos \beta) t \mathbf{j} + \left( (v_0 \sin \beta) t - \frac{g}{2} t^2 \right) \mathbf{k},$$

or, in components,

$$y = (v_0 \cos \beta) t, \quad z = (v_0 \sin \beta) t - \frac{g}{2} t^2. \tag{6.2}$$

The equation for the incline (which is a line in the $y - z$ plane) is given by:

$$z = y \tan \alpha. \tag{6.3}$$

Substituting (6.2) into (6.3), it follows that the projectiles path and the incline intersect at those values of $t$ for which:

$$(v_0 \sin \beta) t - \frac{g}{2} t^2 = ((v_0 \cos \beta) t) \tan \alpha,$$

i.e.

$$t = 0, \quad \text{or} \quad t = \frac{2v_0 (\sin \beta \cos \alpha - \cos \beta \sin \alpha)}{g \cos \alpha} = \frac{2v_0 \sin(\beta - \alpha)}{g \cos \alpha}.$$

The value $t = 0$ gives the intersection point $O$. The second value of $t$ gives point $A$, which is the required point. Using this value of $t$ in the first equation of (6.2), the range of the projectile up the incline is given by:

$$R = y \sec \alpha = (v_0 \cos \beta) \left( \frac{2v_0 \sin(\beta - \alpha)}{g \cos \alpha} \right) \sec \alpha = \frac{2v_0^2 \sin(\beta - \alpha) \cos \beta}{g \cos^2 \alpha}.$$

**Problem 3.** An object slides on a surface along the horizontal straight line $OA$, see figure below.

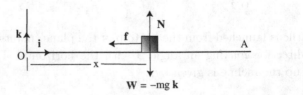

We assume that $x = 0$ and $v = v_0$ at $t = 0$. Suppose that the object comes to rest after traveling a distance $x_0$. Show that the coefficient of friction is:

$$\frac{v_0^2}{2gx_0}.$$

**Solution.** Three forces are acting on the object: the weight, $\mathbf{W} = -mg\mathbf{k}$, the normal force $\mathbf{N}$ of the surface on the object, and the frictional force $\mathbf{f}$. Hence, Newton's equations have the form:

$$m\frac{dv}{dt}\mathbf{i} = \mathbf{W} + \mathbf{N} + \mathbf{f}.$$

But $\mathbf{N} = -\mathbf{W}$, and the magnitude of $\mathbf{f}$ is $f = \mu N = \mu mg$ so that $\mathbf{f} = -\mu mg\mathbf{i}$. Then Newton's equations are written as:

$$m\frac{dv}{dt}\mathbf{i} = -\mu mg\mathbf{i}, \quad \text{or} \quad \frac{dv}{dt} = -\mu g. \tag{6.4}$$

Integrating this equation, and using $v = v_0$ at $t = 0$ gives:

$$v = v_0 - \mu gt, \quad \text{or} \quad \frac{dx}{dt} = v_0 - \mu gt. \tag{6.5}$$

Integrating again, using $x = 0$ at $t = 0$ gives:

$$x = v_0 t - \frac{1}{2}\mu gt^2. \tag{6.6}$$

From (6.5), we see that the object comes to rest (i.e. $v = 0$) when:

$$v_0 - \mu gt = 0 \quad \text{or} \quad t = \frac{v_0}{\mu g}.$$

Substituting this time into (6.6), and noting that $x = x_0$ at this time gives:

$$x_0 = \frac{v_0^2}{\mu g} - \frac{1}{2}\mu g\left(\frac{v_0}{\mu g}\right)^2,$$

or

$$\mu = \frac{v_0^2}{2gx_0}.$$

**Problem 4.** This question is concerned with the first example, the projectile problem, from Chapter 6. The expression for the vertical displacement of the particle as a function of time is given by:

$$z(t) = (v_0 \sin\alpha)\, t - \frac{g}{2}t^2.$$

Clearly, for $t$ sufficiently large, $z(t)$ can be negative. Does this pose any problems?

**Solution.** $z(t)$ negative is a perfectly valid solution of the differential equation governing the dynamics of the projectile. However, a difficulty arises if we want to use the differential equation to model a particular physical situation. For example, if $z = 0$ is the ground (i.e. the "flat Earth") then we cannot consider situations in which $z(t)$ becomes negative.

**Problem 5.** The set-up for this question is the same as the first example, the projectile problem, from Chapter 6, *except* at a distance $d$ from the launch point of the projectile, the horizontal boundary suddenly "drops" to a distance $H$ below zero, see figure below (think of launching the projectile from a cliff).

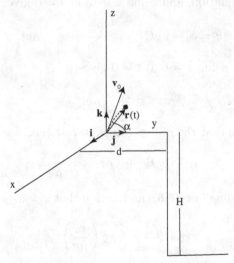

Compute the time that it takes the projectile to reach the vertical distance $H$ below the launch point.

**Solution.** Using the expression for the position of $z$ as a function of time derived in Chapter 6, we have:

$$-H = (v_0 \sin \alpha)t - \frac{g}{2}t^2,$$

or

$$t^2 - \frac{2v_0 \sin \alpha}{g}t - \frac{2H}{g} = 0.$$

Solving this quadratic equation for $t$ gives:

$$t = \frac{v_0 \sin \alpha}{g} \pm \frac{1}{2}\sqrt{\frac{4v_0^2 \sin^2 \alpha}{g^2} + \frac{8H}{g}}.$$

Of the two roots, we take the "plus sign" since the other is negative (and the minus sign, which is perfectly valid from the point of view of the differential equation, is not valid for the physical situation we are modeling):

$$t = \frac{v_0 \sin \alpha}{g} + \sqrt{\frac{v_0^2 \sin^2 \alpha}{g^2} + \frac{2H}{g}}.$$

Now there is a detail we need to check. If the projectile is to go over the "side of the cliff" (and therefore hit the bottom at $z = -H$), the horizontal distance that it travels must be larger than $d$.

**Problem 6.** The set-up for this question is the same as the first example, the projectile problem, from Chapter 6, *except* at a distance $d$ from the launch point of the projectile, a wall is placed of height $H$, see figure below (the wall is parallel to the $x - z$ plane).

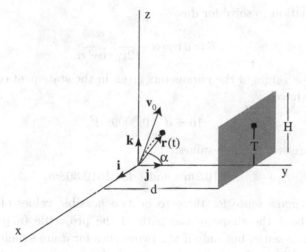

(a) Compute the time that it takes the projectile to hit the wall.
(b) Show that a necessary condition for a particle to hit the wall at a height $T > 0$ is:

$$d < \frac{2v_0^2 \sin \alpha \cos \alpha}{g}.$$

(c) Now suppose $\alpha = 45^0$, $v_0 = 100\,\text{m/s}$, and $g = 9.8\,\text{m/s}^2$. With these parameters fixed, what value of $d$ must we take in order to hit the wall at a height of $10\,\text{m}$?

**Solution.**

(a) $t = \frac{d}{v_0 \cos \alpha}$.

(b) The height that it reaches after this time is:

$$T = d \tan \alpha - \frac{gd^2}{2v_0^2 \cos^2 \alpha}.$$

For the correct physical interpretation, the right-hand side of this expression must be positive, i.e. we must have

$$\tan \alpha > \frac{gd}{2v_0^2 \cos^2 \alpha},$$

or

$$d < \frac{2v_0^2 \sin \alpha \cos \alpha}{g}.$$

(c) The equation to solve for $d$ is:

$$T = d \tan \alpha - \frac{gd^2}{2v_0^2 \cos^2 \alpha}.$$

Using the values of the parameters given in the statement of the problem, we have:

$$10 = d - 0.00098 \, d^2,$$

There are two possible values:

$$d = 10.2m \quad \text{and} \quad d = 1010.308m.$$

Does it make sense for there to be two possible values of $d$? If you think about the shape of the path of the projectile (a parabola) it does make sense, but only if the two values for $d$ are smaller than the range (without the wall being present). Using the parameters given, and the formula for the range from the lecture notes, we compute that $R = 1020.4$m.

# Chapter 7

# Solutions for Problem Set 7

**Problem 1.** Suppose $\mathbf{F}$ is the net force acting on a particle of constant mass $m$, and suppose that $\mathbf{F}$ gives the particle a displacement $d\mathbf{r}$. Then, we defined the work done by $\mathbf{F}$ in moving the particle through a displacement $d\mathbf{r}$ as $dW = \mathbf{F} \cdot d\mathbf{r}$. Recall that Newton's second law is $\mathbf{F} = m\frac{d^2\mathbf{r}}{dt^2}$, i.e. $\mathbf{F}$ is proportional to the second derivative of the displacement $\mathbf{r}$ of the particle. Then shouldn't $\mathbf{F}$ be proportional to the displacement $\mathbf{r}$?

**Solution.** $\mathbf{F}$ and the displacement, $\mathbf{r}$ would be proportional, i.e. lie along the same line, if $\mathbf{r}$ and $\frac{d^2\mathbf{r}}{dt^2}$ were proportional. However, we know that this is not *generally* the case (although it could be true in special cases, see Problem 4 below).

**Problem 2.** Let $\mathbf{r}$ denote the position vector of a particle of constant mass $m$ and let $\mathbf{v} = \frac{d\mathbf{r}}{dt}$ denote the velocity of the particle. Let $\mathbf{A}$ denote a vector such that $\mathbf{F} = \mathbf{v} \times \mathbf{A}$ is the net force acting on the particle. Show that this force does no work on the particle.

**Solution.** A force of this particular form does no work since it is perpendicular to the velocity (think about this in the context of the question above).

**Problem 3.** Explain the following statements:

(a) Work is equal to the transference of kinetic energy.
(b) There can be no work without motion.

45

**Solution.** An issue with both of these questions is how to translate "common language" into mathematical formulae.

(a) We have proven that the work done by the net forces acting on a particle of constant mass $m$ in moving a particle from a point $P_1$ to a point $P_2$ is the kinetic energy of the particle at $P_2$ minus the kinetic energy of the particle at $P_1$.

(b) If we equate motion to nonzero velocity then if there is no motion, there is no velocity (of the particle), and therefore it has no kinetic energy, and therefore no change in kinetic energy is possible.

**Problem 4.** A particle of mass $m$ moves in the $x - y$ plane so that its position vector is:

$$\mathbf{r} = a\cos\omega t\mathbf{i} + b\sin\omega t\mathbf{j},$$

where $a$, $b$, and $\omega$ are positive constants with $a > b$, see figure below.

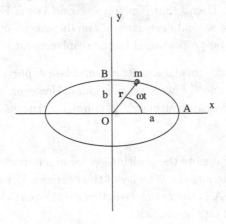

(a) Show that the particle moves on an ellipse.

(b) Show that the force acting on the particle is always directed towards the origin.

(c) Find the kinetic energy of the particle at points $A$ and $B$.

(d) Find the work done by the force field in moving the particle from point $A$ to point $B$ by computing the appropriate line integral.

(e) Using the previous two results, show that the work done in moving from point $A$ to point $B$ is the kinetic energy at $B$ minus the kinetic energy at $A$.

(f) Show that the total work done by the force field in moving the particle once around the ellipse is zero.

(g) Show that the force field is conservative.

(h) Find the potential energy at points $A$ and $B$.

(i) Compute the work done in moving the particle from point $A$ to point $B$ by taking the difference in the potential energy, and show that it is the same as the answer you got in part (d) above.

**Solution.**

(a) The position vector is:

$$\mathbf{r} = x\mathbf{i} + y\mathbf{j} = a\cos\omega t\mathbf{i} + b\sin\omega t\mathbf{j},$$

or

$$x = a\cos\omega t, \quad y = b\sin\omega t,$$

which are just the parametric equations of an ellipse having semi-major axis of length $a$ and semi-minor axis of length $b$. Alternately, since

$$\left(\frac{x}{a}\right)^2 + \left(\frac{y}{b}\right)^2 = \cos^2\omega t + \sin^2\omega t = 1,$$

we also obtain the "other" equation for an ellipse that we usually learn:

$$\frac{x^2}{a^2} + \frac{y^2}{b^2} = 1.$$

(b) Assuming that the particle has constant mass, the force acting on it is:

$$\mathbf{F} = m\frac{d^2\mathbf{r}}{dt^2} = m\frac{d^2}{dt^2}\left(a\cos\omega t\mathbf{i} + b\sin\omega t\mathbf{j}\right),$$

$$= m\left(-\omega^2 a\cos\omega t\mathbf{i} - \omega^2 b\sin\omega t\mathbf{j}\right),$$

$$= -m\omega^2\left(a\cos\omega t\mathbf{i} + b\sin\omega t\mathbf{j}\right) = -m\omega^2\mathbf{r},$$

from which it follows immediately that the force is directed towards the origin.

(c) The velocity is given by:

$$\mathbf{v} = -\omega a\sin\omega t\mathbf{i} + \omega b\cos\omega t\mathbf{j}.$$

Therefore, the kinetic energy is given by:

$$\frac{1}{2}m\mathbf{v}\cdot\mathbf{v} = \frac{1}{2}m\left(\omega^2 a^2\sin^2\omega t + \omega^2 b^2\cos^2\omega t\right).$$

So we have:

Kinetic energy at A (where $\cos \omega t = 1$, $\sin \omega t = 0$) $= \dfrac{1}{2} m \omega^2 b^2$.

Kinetic energy at B (where $\cos \omega t = 0$, $\sin \omega t = 1$) $= \dfrac{1}{2} m \omega^2 a^2$.

(d)          Work done $= \displaystyle\int_A^B \mathbf{F} \cdot d\mathbf{r}$,

$$= \int_0^{\frac{\pi}{2\omega}} \left( -m\omega^2 \left( a \cos \omega t \mathbf{i} + b \sin \omega t \mathbf{j} \right) \right)$$

$$\cdot \left( -\omega a \sin \omega t \mathbf{i} + \omega b \cos \omega t \mathbf{j} \right) dt,$$

$$= \int_0^{\frac{\pi}{2\omega}} m\omega^3 (a^2 - b^2) \sin \omega t \cos \omega t dt,$$

$$= \frac{1}{2} m\omega^2 (a^2 - b^2) \sin^2 \omega t \Big|_0^{\frac{\pi}{2\omega}} = \frac{1}{2} m\omega^2 (a^2 - b^2).$$

(e) Using the previous two results:

$$\text{Work done} = \frac{1}{2} m\omega^2 (a^2 - b^2) = \frac{1}{2} m\omega^2 a^2 - \frac{1}{2} m\omega^2 b^2,$$

$$= \text{kinetic energy at B} - \text{kinetic energy at A}.$$

(f) Using the result above from (d), in making a complete circuit around the ellipse we go from $t = 0$ to $t = \frac{2\pi}{\omega}$. Therefore:

$$\text{Work done} = \int_0^{\frac{2\pi}{\omega}} m\omega^3 (a^2 - b^2) \sin \omega t \cos \omega t dt,$$

$$= \frac{1}{2} m\omega^2 (a^2 - b^2) \sin^2 \omega t \Big|_0^{\frac{2\pi}{\omega}} = 0.$$

(g) The force was obtained in (b). A direct calculation shows that $\nabla \times \mathbf{F} = 0$.

(h) Since the force is conservative there exists a potential $V$ such that:

$$\mathbf{F} = -m\omega^2 x \mathbf{i} - m\omega^2 y \mathbf{j} = -\nabla V = -\frac{\partial V}{\partial x} \mathbf{i} - \frac{\partial V}{\partial y} \mathbf{j} - \frac{\partial V}{\partial z} \mathbf{k}.$$

Then we have:

$$m\omega^2 x = \frac{\partial V}{\partial x}, \quad m\omega^2 y = \frac{\partial V}{\partial y}, \quad \frac{\partial V}{\partial z} = 0.$$

Solving these equations (and setting the integration constant to zero) gives the potential:

$$V = \frac{1}{2}m\omega^2 x^2 + \frac{1}{2}m\omega^2 y^2 = \frac{1}{2}m\omega^2(x^2 + y^2) = \frac{1}{2}m\omega^2 r^2.$$

(i)     Potential at point A (where r=a) $= \frac{1}{2}m\omega^2 a^2,$

Potential at point B (where r=b) $= \frac{1}{2}m\omega^2 b^2.$

Then we have

Work done from A to B = Potential at A − Potential at B,

$$= \frac{1}{2}m\omega^2 a^2 - \frac{1}{2}m\omega^2 b^2,$$

which agrees with the result obtained in (d).

**Problem 5.** Consider the first example in Chapter 6 (the projectile problem). Compute the work done by the net force on the projectile from the launch point to the highest point of the trajectory by:

(a) using the definition of work given by the line integral,
(b) using the definition of work given by the difference of kinetic energies at the two points.

(You should get the same answer for each.)

**Solution.** First we collect together the relevant results from the example that have already been computed in the notes.

$$\mathbf{F} = -mg\mathbf{k},$$

$$\mathbf{v} = v_0 \cos\alpha \mathbf{j} + (v_0 \sin\alpha - gt)\mathbf{k},$$

$$\mathbf{r} = (v_0 \cos\alpha)t\mathbf{j} + \left((v_0 \sin\alpha)t - \frac{1}{2}gt^2\right)\mathbf{k},$$

and, if the particle is launched at $t = 0$, the time required for the particle to reach its highest point is:

$$t_h = \frac{v_0 \sin\alpha}{g}.$$

(a) We have:

$$\int_0^{t_h} \mathbf{F} \cdot d\mathbf{r} = \int_0^{t_h} \mathbf{F} \cdot \frac{d\mathbf{r}}{dt} dt$$

$$= \int_0^{t_h} -mg(v_0 \sin \alpha - gt) dt,$$

$$= -(mg \sin \alpha)t + \frac{mg^2}{2} t^2 \Big|_0^{\frac{v_0 \sin \alpha}{g}},$$

$$= -\frac{m}{2} v_0^2 \sin^2 \alpha.$$

(b) Let $P_1$ denote the point where the projectile is launched (the origin) and $P_2$ denote the highest point of the projectile. Then we have:

$$T_{P_2} - T_{P_1} = \frac{1}{2} m v_0^2 \cos^2 \alpha - \frac{1}{2} m v_0^2 = -\frac{1}{2} m v_0^2 \sin^2 \alpha.$$

**Problem 6.** Consider Problem 1 from the Chapter 6 problems (the inclined plane problem, assuming it to have length $L$). Compute the work done by the net force on the particle resulting from the motion from the top of the incline, to the bottom:

(a) using the definition of work given by the line integral,
(b) using the definition of work given by the difference of kinetic energies at the two points.

(You should get the same answer for each.)

**Solution.** First we collect together the relevant results from the Problem Set 6 Solutions (Problem 1):

$$\mathbf{F} = (mg \sin \alpha - \mu mg \cos \alpha) \mathbf{e_1},$$

$$\mathbf{v} = (g \sin \alpha - \mu g \cos \alpha) t \, \mathbf{e_1},$$

$$s = \frac{g}{2} (\sin \alpha - \mu \cos \alpha) t^2,$$

The particle starts at rest from the top of the incline. If the incline is of length $L$, then the time to reach the bottom is obtained by solving:

$$L = \frac{g}{2} (\sin \alpha - \mu \cos \alpha) t^2,$$

or

$$t_b = \sqrt{\frac{2L}{g(\sin \alpha - \mu \cos \alpha)}}.$$

(a) We have:

$$\int_0^{t_b} \mathbf{F} \cdot d\mathbf{r} = \int_0^{t_b} \mathbf{F} \cdot \frac{d\mathbf{r}}{dt} dt,$$

$$= \int_0^{t_b} mg^2 \left(\sin\alpha - \mu\cos\alpha\right)^2 t\, dt,$$

$$= \frac{mg^2}{2} \left(\sin\alpha - \mu\cos\alpha\right)^2 \left(\sqrt{\frac{2L}{g(\sin\alpha - \mu\cos\alpha)}}\right)^2,$$

$$= mg \left(\sin\alpha - \mu\cos\alpha\right) L.$$

(b) The particle starts from rest at the top of the incline, so at $t = 0$ we have $T_{\text{top}} = 0$. At the bottom of the incline the velocity is given by:

$$\mathbf{v}_b = (g\sin\alpha - \mu g\cos\alpha)\sqrt{\frac{2L}{g(\sin\alpha - \mu\cos\alpha)}}\, \mathbf{e}_1$$

Then the kinetic energy at the bottom of the incline, $T_b$, is given by:

$$T_b = \frac{1}{2}m \left((g\sin\alpha - \mu g\cos\alpha)\sqrt{\frac{2L}{g(\sin\alpha - \mu\cos\alpha)}}\right)^2$$

$$= mg \left(\sin\alpha - \mu\cos\alpha\right) L.$$

# Chapter 8
# Solutions for Problem Set 8

**Problem 1.** Consider a projectile of constant mass $m$ that is launched with initial speed $v_0$ at an angle $\alpha$ with the horizontal. Using conservation of energy, compute:

(a) the maximum height reached,
(b) the position vector at any time.

For an illustration of the geometry see figure below.

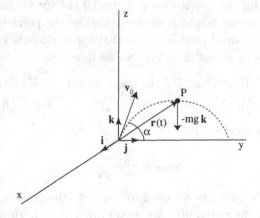

**Solution.** First, we write down some preliminary quantities. The initial velocity is given by:

$$\mathbf{v}_0 = v_0 \cos\alpha\mathbf{j} + v_0 \sin\alpha\mathbf{k}. \tag{8.1}$$

We will denote the (unknown) time-dependent position and velocity vectors by:

$$\mathbf{r}(t) = y(t)\mathbf{j} + z(t)\mathbf{k}, \qquad (8.2)$$

$$\mathbf{v}(t) = \dot{y}(t)\mathbf{j} + \dot{z}(t)\mathbf{k}. \qquad (8.3)$$

Now, here is an important point to understand. The only force acting on the projectile is the gravitational force in the vertical direction, i.e. there are *no* forces in the horizontal direction. Now recall *Newton's First Law:*

*Every particle persists in a state of rest or of uniform motion in a straight line (i.e. with constant velocity) unless acted upon by a force.*

It follows from this that the horizontal component of velocity does not change in time. Hence,

$$\dot{y} = v_0 \cos \alpha.$$

This can be integrated immediately (using $y(0) = 0$) to give:

$$y(t) = (v_0 \cos \alpha)t.$$

Now we return to finding the maximum height reached. We choose the reference point for the gravitational potential energy so that it is zero for $z = 0$. The maximum height is characterized as the point where zero vertical velocity is attained (and we know that the horizontal velocity component is the same as it was initially). So we have:

P.E. at O + K. E. at O = P.E. at max. height + K. E. at max. height

$$0 \quad + \quad \tfrac{1}{2}mv_0^2 \quad = \quad mgz_{max} \quad + \quad \tfrac{1}{2}m(v_0 \cos \alpha)^2,$$

So, a bit of easy algebra gives:

$$z_{max} = \frac{v_0^2 \sin^2 \alpha}{2g}.$$

Now to finish the problem we need to compute the position vector. Since we have already computed $y(t)$, we need only find $z(t)$. How do we do this using energy? The same way. The total energy at an arbitrary point along the path is $\frac{1}{2}m(\dot{y}^2 + \dot{z}^2) + mgz$. Since energy is conserved we can equate this to the total energy at the origin:

$$\frac{1}{2}m(\dot{y}^2 + \dot{z}^2) + mgz = \frac{1}{2}mv_0^2.$$

Substituting $\dot{y}$ from above gives:

$$\frac{1}{2}m((v_0 \cos \alpha)^2 + \dot{z}^2) + mgz = \frac{1}{2}mv_0^2.$$

This gives us an equation for $\dot{z}$ which we can integrate to get $z(t)$:

$$\dot{z} = \sqrt{v_0^2 \sin^2 \alpha - 2gz},$$

or

$$\int_0^z \frac{dz'}{\sqrt{v_0^2 \sin^2 \alpha - 2gz'}} = \int_0^t dt' = t.$$

This integral is relatively straightforward:

$$\int_0^z \frac{dz'}{\sqrt{v_0^2 \sin^2 \alpha - 2gz'}} = \frac{-2\sqrt{v_0^2 \sin^2 \alpha - 2gz'}}{2g}\Bigg|_0^z = t.$$

Working through the algebra, you will find:

$$z = (v_0 \sin \alpha)t - \frac{1}{2}gt^2.$$

**Problem 2.** A particle $P$ of constant mass $m$ slides without rolling down a frictionless inclined plane of angle $\alpha$ and length $\ell$, see figure below.

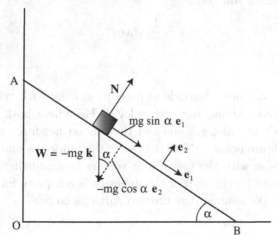

If the particle starts from rest at the top of the incline (at point $A$) use conservation of energy to determine the following quantities:

(a) the velocity as a function of distance traveled,

(b) the distance traveled after time $t$.

(c) Compute the magnitude of the acceleration.

**Solution.** The energy at the top of the incline is solely potential energy given by:

$$mg\ell \sin \alpha.$$

The energy of the particle at an arbitrary point on the incline is:

$$\frac{1}{2}m\dot{s}^2 + mg(\ell - s)\sin \alpha.$$

Equating these (because energy is conserved) gives:

$$mg\ell \sin \alpha = \frac{1}{2}m\dot{s}^2 + mg(\ell - s)\sin \alpha.$$

Hence,

$$\dot{s} = \sqrt{2gs \sin \alpha}.$$

This can be integrated to find $s(t)$:

$$\int_0^s \frac{ds'}{2s' \sin \alpha} = \frac{2s'}{\sqrt{2gs' \sin \alpha}}\bigg|_0^s = \int_0^t dt' = t.$$

After some algebra you get:

$$s = \frac{1}{2}g \sin \alpha t^2.$$

**Problem 3.** Consider a particle of mass $m$ moving with velocity $v_0$ and a particle of mass $M$ moving with velocity $V_0$, where both particles are constrained to move along a line and there is no net force acting on the particles, see figure below. (Note that since the motion is one-dimensional, we are dispensing with the use of unit vectors in expressing the vectorial nature of velocity — the sign of the quantity is adequate for determining the direction.) We assume that the two particles collide.

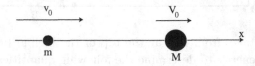

(a) Relate the momentum of the two particles before the collision to the momentum of the two particles after the collision.

(b) Let $x$ denote the position of the particle of mass $m$ and let $X$ denote the position of the particle of mass $M$. We define the *center of mass* of the system of two particles by:

$$\xi = \frac{mx + MX}{m + M}.$$

Show that the velocity of the center of mass, i.e. $\dot{\xi}$, is constant.

(c) An *elastic collision* is defined as a collision where the sum of the kinetic energy of the two particles is the same before and after the collision, i.e.

$$\frac{1}{2}mv_0^2 + \frac{1}{2}MV_0^2 = \frac{1}{2}mv^2 + \frac{1}{2}MV^2. \tag{8.12}$$

Using this result, and the result from part (a), show that the velocities before and after the collision are related as follows:

$$v = \frac{m - M}{m + M}v_0 + \frac{2M}{m + M}V_0,$$

$$V = \frac{M - m}{m + M}V_0 + \frac{2m}{m + M}v_0 \tag{8.13}$$

(d) Consider the case of an elastic collision with $m = M$. What are $v$ and $V$?

(e) Consider the case of an elastic collision with $M$ "very large compared to $m$". Determine $v$ and $V$ in general, and describe the motion in the special case that $V_0 = 0$.

**Solution.**

(a) Since there is no net force acting on the particles we use conservation of momentum and conclude that:

$$mv_0 + MV_0 = mv + MV. \tag{8.4}$$

This answer is correct. However, we are overlooking a point that deserves further thought. Certainly when the particles are not in contact, by assumption, there are no forces acting on either particle. But what about at the instant of contact? Certainly each particle exerts a force on the other (or else the momentum of each particle could not change, even though the total momentum is unchanged). The key here is to consider Newton's third law of motion.

(b) Using (8.4), we have:

$$\dot{\xi} = \frac{m\dot{x} + M\dot{X}}{m + M} = \frac{mv + MV}{m + M} = \frac{mv_0 + MV_0}{m + M} = \text{constant}.$$

(c) We will use momentum conservation and energy conservation to obtain two (linear) equations to solve for the two unknowns $v$ and $V$. We first rewrite

$$\frac{1}{2}mv_0^2 + \frac{1}{2}MV_0^2 = \frac{1}{2}mv^2 + \frac{1}{2}MV^2. \tag{8.5}$$

as

$$m(v^2 - v_0^2) = M(V_0^2 - V^2), \tag{8.6}$$

and (8.4) as:

$$m(v - v_0) = M(V_0 - V). \tag{8.7}$$

Dividing (8.6) by (8.7) gives:

$$v + v_0 = V_0 + V. \tag{8.8}$$

Therefore (8.4) and (8.8) give the following pair of equations to solve for the unknowns $v$ and $V$:

$$mv + MV = mv_0 + MV_0,$$
$$v - V = -v_0 + V_0. \tag{8.9}$$

These two equations can easily be solved for $v$ and $V$ to obtain the result:

$$v = \frac{m - M}{m + M}v_0 + \frac{2M}{m + M}V_0,$$
$$V = \frac{M - m}{m + M}V_0 + \frac{2m}{m + M}v_0. \tag{8.10}$$

(d) Using (8.10), we see that for $m = M$ it follows that $v = V_0$ and $V = v_0$. In other words, after the collision the mass on the left moves with the initial velocity of the mass on the right, and the mass on the right moves with the initial velocity of the mass on the left.

(e) We compute the limit as $M \to \infty$ to obtain:

$$v = -v_0 + 2V_0,$$
$$V = V_0. \tag{8.11}$$

If $V_0 = 0$ we see that the small mass "bounces off" the large mass and reverses its direction with the same speed, but velocity is in the opposite direction.

# Chapter 9

# Solutions for Problem Set 9

**Problem 1.** Consider the equation:

$$\ddot{s} = -s,$$

where, for simplicity, we have set $m = 1$ (remember from an earlier home-work problem that we can "rescale time" so that the mass becomes unity).

(a) Write it as a first order system, or vector field, on the phase plane.
(b) Compute the potential energy and sketch it.
(c) Find all equilibria and classify their stability.
(d) Sketch the phase portrait.
(e) Compute expressions for the trajectories in the phase plane as a function of time (and take $s(0) = 0$ for simplicity).

**Solution.**

(a)
$$\dot{s} = v,$$
$$\dot{v} = -s.$$

(b) $V(s) = \frac{s^2}{2}$ sketched in Fig. 9.1.
(c) Stable equilibria at $(s, v) = (0, 0)$ (relative minima of the potential).
(d) Phase portrait sketched in Fig. 9.1.
(e) Using the expression derived from the energy integral in class:

$$\int_0^s \frac{ds'}{\sqrt{2E - s'^2}} = \left. \sin^{-1} \frac{s'}{\sqrt{2E}} \right|_0^s = t.$$

or

$$s = \sqrt{2E} \sin t, \quad \text{and} \quad v = \dot{s}.$$

59

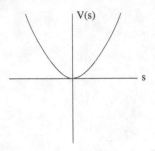

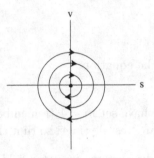

Fig. 9.1. Graph of the potential energy function and the phase portrait.

**Problem 2.** Consider the equation:

$$\ddot{s} = s - s^3,$$

where, for simplicity, we have set $m = 1$.

(a) Write it as a first order system, or vector field, on the phase plane.
(b) Compute the potential energy and sketch it.
(c) Find all equilibria and classify their stability.
(d) Sketch the phase portrait.

**Solution.**

(a)
$$\dot{s} = v,$$

$$\dot{v} = s - s^3.$$

(b) $V(s) = -\frac{s^2}{2} + \frac{s^4}{4}$ sketched in Fig. 9.2.

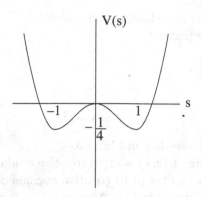

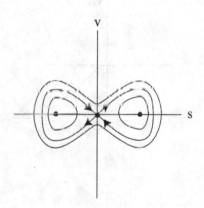

Fig. 9.2. Graph of the potential energy function and the phase portrait.

(c) Stable equilibria at $(s, v) = (\pm 1, 0)$ (relative minima of the potential),
a saddle point at $(s, v) = (0, 0)$ (relative maxima of the potential).
(d) Phase portrait sketched in Fig. 9.2.

**Problem 3.** Consider the equation:
$$\ddot{s} = s - s^2,$$
where, for simplicity, we have set $m = 1$.

(a) Write it as a first order system, or vector field, on the phase plane.
(b) Compute the potential energy and sketch it.

(c) Find all equilibria and classify their stability.

(d) Sketch the phase portrait.

**Solution.**

(a)
$$\dot{s} = v,$$
$$\dot{v} = s - s^2.$$

(b) $V(s) = -\frac{s^2}{2} + \frac{s^3}{3}$ sketched in Fig. 9.3.

(c) Stable equilibrium at $(s, v) = (1, 0)$ (relative minima of the potential), a saddle point at $(s, v) = (0, 0)$ (relative maxima of the potential).

(d) Phase portrait sketched in Fig. 9.3.

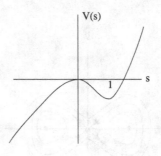

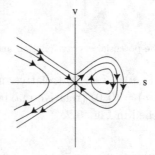

Fig. 9.3. Graph of the potential energy function and the phase portrait.

# Chapter 10

# Solutions for Problem Set 10

Problems 1–7 below are concerned with the motion of a particle of constant mass $m$ in one dimension under the action of a conservative force:

$$\dot{s} = v,$$

$$\dot{v} = -\frac{1}{m}\frac{dV}{ds}(s),$$

where $V(s)$ is the potential function.

**Problem 1.** Show that all level sets of the energy function cross the $s$-axis in the phase plane perpendicularly. (Hint: Compute the slope of the level set at the turning point, $v = 0$.)

**Solution.** We have

$$v = \pm\sqrt{\frac{2}{m}}\sqrt{E - V(s)}.$$

It follows that:

$$\frac{dv}{ds} = \mp\sqrt{\frac{1}{2m\left(E - V(s)\right)}}\frac{dV}{ds}(s).$$

At the crossing point (or, as I referred to it in class, turning point), $v = 0$, $E = V(s)$. However, we need to take into account both branches of this function, i.e. the $\pm$ signs. Look at each branch separately and consider the limiting behavior of the slope as the $s$-axis is approached.

63

The question could also be answered by implict differentiation. A level set of the energy is given by:

$$H(s, v) = \frac{1}{2}mv^2 + V(s) = E.$$

Implictly differentiating with respect to $s$ gives:

$$mv\frac{dv}{ds} + \frac{dV}{ds} = 0,$$

or

$$\frac{dv}{ds} = -\frac{1}{mv}\frac{dV}{ds}$$

which is infinite at $v = 0$. (In order to draw this conclusion, do we have to say something about $\frac{dV}{ds}$ at the turning point?)

Make sure you understand what it means, and what we are assuming, when we implicitly differentiate the energy function. If you don't, ask for an explanation.

**Problem 2.** Suppose you add a constant (i.e. a real number) to the potential function. How does the corresponding phase portrait change?

**Solution.** The phase portrait does not change at all in the sense of the geometry of the level sets of the energy function. However, the *value* of the energy for the different level sets changes according to the value of the constant.

**Problem 3.** Suppose $(s, v) \equiv (s_0, 0)$ is an equilibrium point of Newton's equations above. Is it a solution of Newton's equations?

**Solution.** Yes. Substitute this (constant) function into the left-hand side, and right-hand side, of Newton's equations and show equality. What do the left- and right-hand sides equal?

**Problem 4.** Consider the potential energy sketched in the figure below.

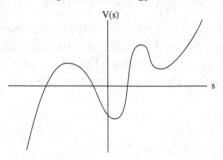

(a) Determine the number of equilibria and their stability type.

(b) Sketch the phase portrait.

**Solution.** See Fig. 10.1. There are four equilibria: two stable and two unstable.

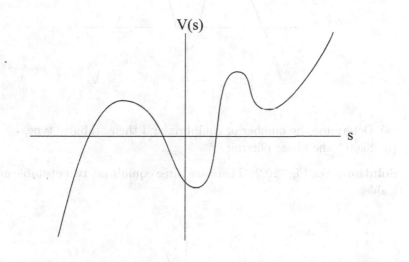

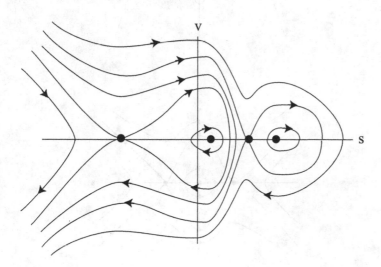

Fig. 10.1. Graph of the potential energy function and the phase portrait.

**Problem 5.** Consider the potential energy sketched in the figure below.

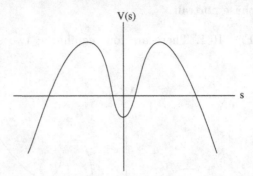

(a) Determine the number of equilibria and their stability type.
(b) Sketch the phase portrait.

**Solution.** See Fig. 10.2. There are three equilibria: two unstable and one stable.

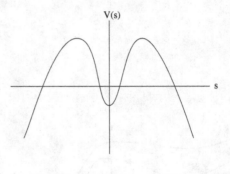

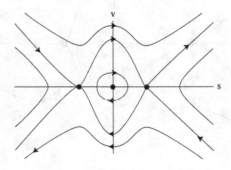

Fig. 10.2. Graph of the potential energy function and the phase portrait.

**Problem 6.** Consider the potential energy sketched in the figure below.

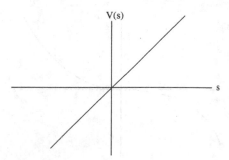

(a) Determine the number of equilibria and their stability type.
(b) Sketch the phase portrait.

**Solution.** See Fig. 10.3. There are no equilibria.

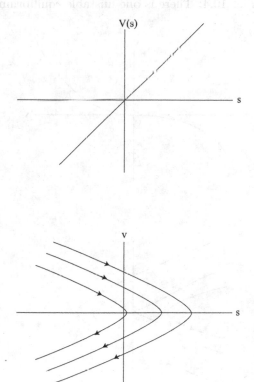

Fig. 10.3. Graph of the potential energy function and the phase portrait.

**Problem 7.** Consider the potential energy sketched in the figure below.

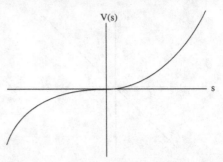

There is an inflection point at the origin (i.e. think of $V(s) = s^3$).

(a) Determine the number of equilibria and their stability type.
(b) Sketch the phase portrait.

**Solution.** See Fig. 10.4. There is one unstable equilibrium.

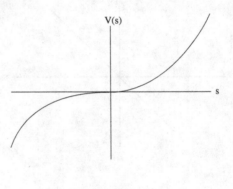

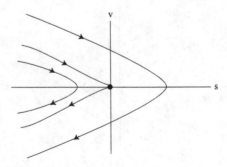

Fig. 10.4. Graph of the potential energy function and the phase portrait.

**Problem 8.** The angular momentum (about some point) as a function of time is given by:

$$\Omega = 6t^2\mathbf{i} - (2t+1)\mathbf{j} + (12t^3 - 8t^2)\mathbf{k}.$$

Find the torque (about the same point) at $t = 1$.

**Solution.**

$$\Lambda = \frac{d\Omega}{dt} = 12t\mathbf{i} - 2\mathbf{j} + (36t^2 - 16t)\mathbf{k}.$$

evaluating this expression at $t = 1$ gives:

$$12\mathbf{i} - 2\mathbf{j} + 20\mathbf{k}.$$

**Problem 9.** A particle of mass 2 moves in a time-dependent force field given by:

$$\mathbf{F} = 24t^2\mathbf{i} + (36t - 16)\mathbf{j} - 12t\mathbf{k}.$$

Assume that at $t = 0$ the particle has the following position and velocity:

$$\mathbf{r}_0 = 3\mathbf{i} - \mathbf{j} + 4\mathbf{k},$$
$$\mathbf{v}_0 = 6\mathbf{i} + 15\mathbf{j} - 8\mathbf{k}.$$

Compute, for any time $t$:

(a) the velocity,
(b) the position,
(c) the torque about the origin,
(d) the angular momentum about the origin.

**Solution.**

(a) From Newton's second law:

$$2\frac{d\mathbf{v}}{dt} = 24t^2\mathbf{i} + (36t - 16)\mathbf{j} - 12t\mathbf{k},$$

or,

$$\frac{d\mathbf{v}}{dt} = 12t^2\mathbf{i} + (18t - 8)\mathbf{j} - 6t\mathbf{k}.$$

Integrating this expression with resepct to $t$ gives:

$$\mathbf{v} = 4t^3\mathbf{i} + (9t^2 - 8t)\mathbf{j} - 3t^2\mathbf{k} + \mathbf{c}_1.$$

At $t = 0$, $\mathbf{v} = \mathbf{v}_0$, and therefore:

$$\mathbf{v}_0 = 6\mathbf{i} + 15\mathbf{j} - 8\mathbf{k} = \mathbf{c}_1,$$

and we have:

$$\mathbf{v} = (4t^3 + 6)\mathbf{i} + (9t^2 - 8t + 15)\mathbf{j} - (3t^2 + 8)\mathbf{k}.$$

(b) Using the previous result:

$$v = \frac{d\mathbf{r}}{dt} = (4t^3 + 6)\mathbf{i} + (9t^2 - 8t + 15)\mathbf{j} - (3t^2 + 8)\mathbf{k}.$$

Integrating this expression gives:

$$\mathbf{r} = (t^4 + 6t)\mathbf{i} + (3t^3 - 4t^2 + 15t)\mathbf{j} - (t^3 + 8t)\mathbf{k} + \mathbf{c}_2.$$

At $t = 0$, $\mathbf{r} = \mathbf{r}_0$, and therefore:

$$\mathbf{r} = \mathbf{r}_0 = 3\mathbf{i} - \mathbf{j} + 4\mathbf{k} = \mathbf{c}_2,$$

and we have:

$$\mathbf{r} = (t^4 + 6t + 3)\mathbf{i} + (3t^3 - 4t^2 + 15t - 1)\mathbf{j} - (t^3 + 8t - 4)\mathbf{k}.$$

(c)

$$\begin{aligned}
\boldsymbol{\Lambda} = \mathbf{r} \times \mathbf{F} &= \left((t^4 + 6t + 3)\mathbf{i} + (3t^3 - 4t^2 + 15t - 1)\mathbf{j} - (t^3 + 8t - 4)\mathbf{k}\right) \\
&\quad \times \left(24t^2\mathbf{i} + (36t - 16)\mathbf{j} - 12t\mathbf{k}\right), \\
&= \left(32t^3 + 108t^2 - 260t + 64\right)\mathbf{i} - \left(12t^5 + 192t^3 - 168t^2 - 36t\right)\mathbf{j} \\
&\quad - \left(36t^5 - 80t^4 + 360t^3 - 240t^2 - 12t + 48\right)\mathbf{k}
\end{aligned}$$

(d)

$$\begin{aligned}
\boldsymbol{\Omega} = \mathbf{r} \times (m\mathbf{v}) = m(\mathbf{r} \times \mathbf{v}) &= 2\left((t^4 + 6t + 3)\mathbf{i}\right. \\
&\quad + (3t^3 - 4t^2 + 15t - 1)\mathbf{j} - (t^3 + 8t - 4)\mathbf{k}) \\
&\quad \times \left((4t^3 + 6)\mathbf{i} + (9t^2 - 8t + 15)\mathbf{j} - (3t^2 + 8)\mathbf{k}\right), \\
&= (8t^4 + 36t^3 - 130t^2 + 64t - 104)\mathbf{i} \\
&\quad - (2t^6 + 48t^4 - 56t^3 - 18t^2 - 96)\mathbf{j} \\
&\quad - (6t^6 - 16t^5 + 90t^4 - 80t^3 - 6t^2 + 48t - 102)\mathbf{k}
\end{aligned}$$

**Problem 10.** A *central force* is a force having the following form:

$$\mathbf{F} = f(r)\mathbf{r}_1 = f(r)\frac{\mathbf{r}}{r},$$

where $f(r)$ is an arbitrary function of the magnitude of the position vector. Show that angular momentum about the origin is conserved for a particle of constant mass $m$ moving under the influence of a central force field.[1]

---

[1]This is an important result that we will use in the next chapter.

**Solution.** The time rate of change of the angular momentum about the origin is given by the torque about the origin. Therefore we only need to show that the torque about the origin is zero. This is a trivial computation:

$$\mathbf{\Lambda} = \mathbf{r} \times \mathbf{F} = \mathbf{r} \times f(r)\frac{\mathbf{r}}{r} = \frac{f(r)}{r}(\mathbf{r} \times \mathbf{r}) = 0.$$

**Problem 11.** Show that if a particle moves under the influence of a central force field, then its path must always lie in a plane. (Hint. A plane is defined by a constant vector. It then suffices to show that the position vector is perpendicular to an appropriately chosen constant vector. Use the previous problem to choose this constant vector.)[2]

**Solution.** It follows from the previous problem that:

$$\mathbf{r} \times \mathbf{F} = 0,$$

and therefore

$$\mathbf{r} \times m\frac{d\mathbf{v}}{dt} = 0,$$

or

$$\mathbf{r} \times \frac{d\mathbf{v}}{dt} = 0,$$

which is the same as (why?)

$$\frac{d}{dt}(\mathbf{r} \times \mathbf{v}) = 0.$$

Integrating this equation with respect to time gives:

$$\mathbf{r} \times \mathbf{v} = \mathbf{h}, \qquad\qquad (10.1)$$

where $\mathbf{h}$ is a constant vector. Now $\mathbf{r} \times \mathbf{v}$ is perpendicular to $\mathbf{r}$ (why?). Therefore taking the dot product of both sides of (10.1) with $\mathbf{r}$ gives:

$$\mathbf{r} \cdot (\mathbf{r} \times \mathbf{v}) = 0 = \mathbf{r} \cdot \mathbf{h}.$$

Therefore, the position vector is always perpendicular to the constant vector $\mathbf{h}$, so that the motion is always in a plane.

---

[2]This is also an important result that we will use in the next chapter.

# Chapter 11

# Solutions for Problem Set 11

**Problem 1.** Prove that in cartesian coordinates the magnitude of the areal velocity is $\frac{1}{2}(x\dot{y} - y\dot{x})$.

**Solution.** The magnitude of the areal velocity is given by $\frac{1}{2}|\mathbf{r} \times \mathbf{v}|$. Hence, we need to compute $\mathbf{r} \times \mathbf{v}$ in cartesian coordinates, and then compute the magnitude of the resulting vector.

$$\mathbf{r} \times \mathbf{v} = (x\mathbf{i} + y\mathbf{j}) \times (\dot{x}\mathbf{i} + \dot{y}\mathbf{j}) = x\dot{y}\mathbf{k} - y\dot{x}\mathbf{k}.$$

Then

$$|\mathbf{r} \times \mathbf{v}| = x\dot{y} - y\dot{x}.$$

**Problem 2.** Derive the equation:

$$\frac{d^2r}{d\theta^2} - \frac{2}{r}\left(\frac{dr}{d\theta}\right)^2 - r = \frac{r^4 f(r)}{mh^2}.$$

**Solution.** We start with the equation derived in class:

$$\ddot{r} - \frac{h^2}{r^3} = \frac{f(r)}{m}. \tag{11.1}$$

We need two preliminary relations. From $r^2\dot{\theta} = h$ we have:

$$\dot{\theta} = \frac{h}{r^2}. \tag{11.2}$$

Differentiating $r^2\dot{\theta} = h$ with respect to $t$ gives:

$$2r\dot{r}\dot{\theta} + r^2\ddot{\theta} = 0,$$

or

$$\ddot{\theta} = -\frac{2\dot{\theta}}{r}\dot{r} = -\frac{2h}{r^3}\dot{r}. \tag{11.3}$$

Now we use the chain rule:

$$\frac{dr}{dt} = \frac{dr}{d\theta}\frac{d\theta}{dt} = \dot{\theta}\frac{dr}{d\theta} = \frac{h}{r^2}\frac{dr}{d\theta}. \tag{11.4}$$

$$\frac{d^2r}{dt^2} = \left(\frac{d}{dt}\left(\frac{dr}{d\theta}\right)\right)\dot{\theta} + \frac{dr}{d\theta}\ddot{\theta},$$

$$= \frac{d^2r}{d\theta^2}\dot{\theta}^2 + \frac{dr}{d\theta}\ddot{\theta},$$

$$= \frac{h^2}{r^4}\frac{d^2r}{d\theta^2} - \frac{2h^2}{r^5}\left(\frac{dr}{d\theta}\right)^2,$$

where we have used (11.2), (11.3) and (11.4). (11.5)

Now substituting (11.5) into (11.1) gives the result.

**Problem 3.** Show that the position of a particle as a function of time moving in a central force field can be determined from the equations:

$$t = \int \frac{1}{\sqrt{G(r)}}dr, \qquad t = \frac{1}{h}\int r^2 d\theta, \tag{11.19}$$

where

$$G(r) = \frac{2E}{m} + \frac{2}{m}\int f(r)dr - \frac{2h^2}{m^2r^2}.$$

**Solution.** This result uses conservation of energy. From class we derived the following equation that expresses conservation of energy for a particle moving in a central force field:

$$\frac{1}{2}m\left(\dot{r}^2 + r^2\dot{\theta}^2\right) - \int f(r)dr = E.$$

Substituting $\dot{\theta} = \frac{h}{r^2}$ into this equation gives:

$$\frac{1}{2}m\left(\dot{r}^2 + \frac{h^2}{r^2}\right) - \int f(r)dr = E,$$

or

$$\dot{r}^2 = \frac{2E}{m} + \frac{2}{m}\int f(r)dr - \frac{2h^2}{mr^2} \equiv G(r).$$

From this expression we obtain:

$$\frac{dr}{dt} = \sqrt{G(r)},$$

or

$$t = \int \frac{1}{\sqrt{G(r)}} dr.$$

The second equation follows by writing $\dot{\theta} = \frac{h}{r^2}$ as:

$$dt = \frac{1}{h} r^2 d\theta.$$

## Problem 4.

(a) Find the potential energy for a particle which moves in the force field:

$$\mathbf{F} = -\frac{K}{r^2} \mathbf{r}_1,$$

where $K$ is some positive constant.

(b) How much work is done by the force field in moving a particle from a point on the circle of radius $r = a > 0$ to a point on the circle of radius $r = b > 0$? Does the work depend on the path?

**Solution.**

(a) The potential is given by:

$$V(r) = \int \frac{K}{r^2} dr = -\frac{K}{r}.$$

(b) The work done is given by:

$$V(r = a) - V(r = b) = \frac{K}{b} - \frac{K}{a} = \frac{K(a - b)}{ab}.$$

**Problem 5.** Consider the relation:

$$r^2 \dot{\theta} = h = \text{constant},$$

that we derived earlier. Explain how it enables us to determine the $\theta$ component of motion if we know the $r$ component of motion.

**Solution.** From

$$r^2 \dot{\theta} = h = \text{constant},$$

we derive the quadrature:

$$\int d\theta = h \int \frac{dt}{r(t)^2}.$$

**Problem 6.** How is the expression $r^2\dot{\theta}$ related to the angular momentum of the particle about $O$?

**Solution.** In the chapter we showed that:

$$\mathbf{r} \times \mathbf{v} = r^2\dot{\theta}\mathbf{k}.$$

Hence, $mr^2\dot{\theta}\mathbf{k}$ is the angular momentum of the particle about $O$.

Printed in the United States
by Baker & Taylor Publisher Services

Printed in the United States
by Baker & Taylor Publisher Services